Praise for *Mycelial Mayhem*

Mycelial Mayhem is a comprehensive and personal tome on the production, business, promotion, and problem solving for the independent mushroom grower. Although my interests lean toward "mushroom art" more than mushroom production, I am looking forward to having my own copy for its valuable resources.

— Taylor Lockwood

This lively, companionable book demystifies mushrooms for those of us seeking clear explanations and checklists combined with infectious enthusiasm and some stunning photos, to nudge us towards starting out cultivating, wild collecting and marketing this diverse class of crops, which can complement our other gardening and farming ventures.

— Pam Dawling, author, *Sustainable Market Farming*

Usually I'm out in the woods looking for wild mushrooms, not cultivating them, but after reading *Mycelial Mayhem*'s approachable primer to this age-old art I'm ready to invite the mysterious and captivating kingdom of fungi right into my own home.

— Langdon Cook, author, *The Mushroom Hunters*

This is the book that I will recommend for all beginning mushroom growers. Whether you want to grow mushrooms just for your own use or are considering them as a business, you will find this is a practical, down to earth, and enjoyable book to read. It covers everything the beginner needs to know including the basics of mushroom biology, how to grow, wild collecting, incorporating mushrooms into your landscape, cooking, and how start a mushroom business. The niche business and supplement income section is excellent and appropriate for someone considering any kind of small-scale agricultural business.

— Jeanine Davis, PhD., Associate Professor and Extension Specialist
North Carolina State University, and lead author, *Growing and Marketing Ginseng,*
Goldenseal and Other Woodland Medicinals

Mycelial Mayhem draws us into the wild, weird, and wonderful world of the fungal kingdom with irresistible wit and wisdom. Emphasizing culinary glory, responsible foraging, nutritional and health benefits, the role of fungi in healthy ecosystems, and the barely explored possibilities of profitable business, this charming and very well-informed book is both welcome and indispensable for the beginning cultivator or curious cook.

— Peter Bane, author, *The Permaculture Handbook*

MYCELIAL MAYHEM

MYCELIAL MAYHEM

Growing Mushrooms for Fun, Profit and Companion Planting

David and Kristin Sewak

Cover design by Diane McIntosh.
Cover images © iStock

Printed in Canada. First printing May 2016

Funded by the   Financé par le
Government   gouvernement | Canadä
of Canada   du Canada

Inquiries regarding requests to reprint all or part of *Mycelial Mayhem* should be addressed to New Society Publishers at the address below. To order directly from the publishers, please call toll-free (North America) 1-800-567-6772, or order online at www.newsociety.com

Any other inquiries can be directed by mail to:

New Society Publishers
P.O. Box 189, Gabriola Island, BC V0R 1X0, Canada
(250) 247-9737

LIBRARY AND ARCHIVES CANADA CATALOGUING IN PUBLICATION

Sewak, David, author
Mycelial mayhem : growing mushrooms for fun, profit and companion planting / David and Kristin Sewak.

Includes bibliographical references and index.
Issued in print and electronic formats.
ISBN 978-0-86571-814-2 (paperback). — ISBN 978-1-55092-621-7 (ebook)

1. Mushroom culture.   2. Edible mushrooms.   3. Mycelium.   4. Mushrooms.
I. Sewak, Kristin, author   II. Title.

SB353.S47 2016   635'.8   C2015-907723-0
                                                       C2015-907724-9

New Society Publishers' mission is to publish books that contribute in fundamental ways to building an ecologically sustainable and just society, and to do so with the least possible impact on the environment, in a manner that models this vision. We are committed to doing this not just through education, but through action. The interior pages of our bound books are printed on Forest Stewardship Council®-registered acid-free paper that is 100% post-consumer recycled (100% old growth forest-free), processed chlorine-free, and printed with vegetable-based, low-VOC inks, with covers produced using FSC®-registered stock. New Society also works to reduce its carbon footprint, and purchases carbon offsets based on an annual audit to ensure a carbon neutral footprint. For further information, or to browse our full list of books and purchase securely, visit our website at: www.newsociety.com

Dedicated to Papap,
from the old generation of mushroom hunters,
and our children,
Aedan, Cassidy and Sage,
the next generation of mushroom hunters.
Keep spreading the spores.

Contents

Section III: The Fruit of Your Labor

Section IV: Spreading the Spores

Acknowledgments

At times, writing this book has been nothing but mayhem, but we mean that in a good way. Still, without the inspiration and support of many, *Mycelial Mayhem* would not have been possible. We'd like to thank our family and friends for all of their love and support throughout the years. You've collectively made us who we are, so thank you all.

To the people who have inspired our mushroom-growing journey, Papap Martin Vrtis, Nana Judith Vrtis, Dr. Sewak, Violet Sewak, Jody, Kutka, and Hasha Venn of Tsoma Farms—our family and fungi-friends, Joe Krawczyk and Mary Ellen Kozak of Field & Forest Products, Susan, the head chef at Green Gables Restaurant, thank you. Indirectly, we'd like to thank Paul Stamets and the wealth of knowledge you have provided the world, along with David Arora and Sepp Holzer, through your books and examples. We stand on your shoulders and hope we are spreading the spores in rhythm with your visions.

Our appreciation is also owed to our logger friend, Ron Gillingham, for the constant supply of logs he provided to us and for teaching Dave how to cut wood without killing himself. Ron always told Dave, "You cut like a girl in a skirt!"

Thank you, Rigel Richardson, for your assistance and knowledge during your summer mushroom-growing apprenticeship with us, and for being brave enough to leave the big city to stay with us hillbillies in your hand-built yurt. Rigel and Dave perfected the art of "road hunting for mushrooms" by driving around after rains to find wild chicken of the woods and oyster mushrooms in the state forest. Try it, you may be surprised what you find! Thank you, Lindsay Pyle, also one of our mushroom-growing apprentices. Rigel and Lindsay, continue to spread those spores of mushroom knowledge and inspiration.

To Ernie Lee of Keelboat Farms, thank you for your insight with part of the book.

To the team at New Society Publishers, Ingrid Witvoet, Greg Green, Sue Custance, Sara Reeves and EJ Hurst. We'd also like to thank Linda Glass, our copy editor, for helping us make this book a better product. Thank you for the opportunity to share our knowledge.

Dave says, "Without a doubt, thank you, Kristin Sewak (because Dave loves to say he writes like Jack Kerouac, and he does!). Massive bursts of stream-of-consciousness writings that jump around, weave, and bob and with maddening punctuation! Without you, this book would have been a jumble of gibberish. Also for your faith to believe and see, as in the garden in the front yard, the bird bath stump and every other crazy idea that popped into my head—like the chicken coop…but that's a different story for another day!"

Kristin says, "Thanks, Dave Sewak, for your charisma and knowledge, which launched *Mycelial Mayhem*, and for making the journey fun, crazy, and unpredictable!"

Foreword

by Joe Krawczyk

In 1983, my wife and I inoculated our first shiitake logs on what would later become our homestead in Peshtigo, Wisconsin, and the future site of our spawn production business known as Field and Forest Products. My future father-in-law at the time asked incredulously and with some alarm, "You are going to do this for a livelihood?" We only wish he was around today to see how far the shiitake industry has progressed, let alone the interest in cultivation and consumption of other cultivated and wild harvested mushrooms.

Most of us can go into a grocery store and find the classic white button mushroom in the produce section and know that it is safe to eat and really good on pizza. We probably heard at one time or another that they grew on horse manure in caves which really added to the mystery of this mushroom but hopefully not its taste. Then out of the blue came the strange but at least related "gourmet" Portobello and soon on its heels followed Shiitake, otherwise known as the Black or Japanese Forest Mushroom. Oysters, Maitake and other cultivated mushrooms now grace many produce aisles, some with names the average person can't make sense of such as Shiimiji and Enokitake. We often now see edible wild mushrooms offered, even in the middle of winter. How did this come about? And what about all these other mushrooms; do they also grow in caves and on horse manure?

The grocery store mushrooms and those we find outdoors in meadows and the forest are the macrofungi; ones we can see with the naked eye. There are also a lot of macrofungi in the world, with more being discovered each year. It is easy to overlook the ones we cannot see, although these too have their place in changing us and the world we live in.

Fungi are great agents of change in our world. We do not realize this on a daily basis because for the most part, fungi take their time to get things done in a slow, methodical and unheralded manner. It would be hard to imagine a world where decay fungi are not hard at work because woody debris would be piled miles deep on our home world! But yet, on a day-to-day basis, we rarely give it notice, going about our daily routines not paying mind to the course of their tireless and steady work and the many influences they have on us and the environment in which we reside. These influences sometimes may be subtle and can go on unnoticed for a long period of time, while sometimes, they can bring about drastic change on somewhat short order.

One only needs to travel a little back in time to the 1960s and early 70s for a recent example. Many of us can remember the effect that the microscopic fungus Ophiostoma ulmi, or Dutch Elm Disease, had on most urban forests and street tree plantings. In the eastern United States and Canada, neighborhood streets were lined with one species, the American Elm. A stately tree whose vase-shaped form produced a tunnel of green that provided shade for urban dwellers and habitat for numerous songbirds. Within a few decades, that urban forest was greatly altered, decimated by Dutch Elm Disease. This fungal disease was introduced into the United States in 1930 via infested elm logs imported from Europe and was carried long distances from elm to elm via domestic and imported bark beetles and locally via root grafting between closely spaced trees. For the most part, the diseased wood was either chipped or landfilled where beneficial fungi started their job of converting the wood to organic manner. It is a shame that at that time, very few people in North America that were involved with the suppression, control and wood disposal of these diseased trees were not acquainted with the Wine Cap mushroom and its great ability to convert wood chips into organic matter and during that process produce a delightful edible mushroom.

Although it's easy to develop a negative attitude when confronted with disasters caused by fungi that were brought about by our own ignorance of their biology, our knowledge of pathogenic fungi and cultivated mushrooms has grown exponentially since that time. The world has simply become more connected and our population has become more diverse, introducing not only new fungal diseases but edible fungi and new ways to cultivate them and eat them.

Who would have dreamed of an industry based on decaying oak and other hardwood logs to produce a whole new assortment of mushrooms recently unheard of in the North American marketplace? Numerous strains of shiitake and oyster mushrooms, Nameko, Hen of the Woods, Reishi, Lion's Mane, Comb Tooth, Olive Oysterling, Kuratake, Piopini, Enoki, Shiimiji, Wine Cap, are now available to US growers and gardeners. Blewit, a leaf litter decomposer, and the compost grown Almond Agaricus, have also been added to this list. Following the dream of cultivating mushrooms as a hobby or as a small scale enterprise is now within reach of many people. Mushrooms have become a part of the agroforestry quilt of our nation, being produced by backyard growers for personal use and by farmers wanting to add to their CSA offerings and farm market stands.

While there are many claims about the medicinal nature of many of these fungi, growing mushrooms as a business or hobby is good medicine in itself. In terms of physical activity, cutting, hauling and inoculating logs becomes the perfect woodland workout clinic. For instance, preparing a shiitake bed log requires handling it eight separate times from start to finish. Throw in the additional activities of soaking logs, stacking them, and picking the crop, and one is well on their way to better health.

The mental health benefits also need to be taken into account. Growing mushrooms provides a great reason to get outdoors and notice all of the wildlife, plant and animal, residing in the cultivation environment. There is a certain joy to being outdoors on a crisp fall day and firing up the chainsaw to fell trees for mushroom cultivation. For that joy to happen though, the sawyer needs to step back and observe the stand of trees whose future they are about alter. Selecting which tree to cut and what influence its removal will have on its surrounding neighbors is a great mental exercise, as the sawyer will have to consider what the ramifications of his actions will be now, and 50 years down the road.

When growing mushrooms for the first time, one closely resembles an expectant parent. The new grower will find themselves out looking after their mushroom crop, during all sorts of weather conditions, waiting for their arrival. Then, when they finally do start to appear, a smile will erupt across their face followed by a nod of the head for a job well done. They will know it was worth the wait and the agony that went along with it. The pleasure doesn't stop there though! In the ensuing springs and falls, as their logs mature, the highest quality mushrooms are produced, and the pleasure of picking high quality, organically grown mushrooms is one that never gets old. Couple the pleasure of picking the crop with the experience of cooking and eating the freshest mushrooms ever, and life is definitely good.

The whole world of edible wild fungi also has a lot to offer us in the realm of good things to eat and physical and mental well-being. During a walk on a beautiful late summer or early fall day in your favorite wooded area, have you ever wondered what 'that' mushroom is? Could it be edible? Could it kill me? Once your curiosity was satisfied about that mushroom, perhaps your interest flowered to take a more serious look at wild edible mushrooms and when to find them, what trees they grow in association with, and really most importantly, how to cook with them. For some of us, particularly those of eastern European descent, mushroom hunting and general knowledge about mushrooms and their use might be genetically linked. For the rest of us, sources of information from others need to be sought out to help identify and sort through as to what is really worthwhile to hunt, collect and eat. It's a small step to take for a long and fulfilling journey.

—Joseph H Krawczyk
Co-Author "Growing Shiitake in a Continental Climate" 1988
Peshtigo WI
November 21, 2015

Introduction

What is *Mycelial Mayhem?*

Mycelium, noun, plural, mycelia
[mahy-see-lee-uh] Mycology
1. the mass of hyphae that form the vegetative part of a fungus.

Mayhem, noun
[mey-hem, mey-hum]
a state of rowdy disorder

So, you're wondering: "This book is about the rowdy disorder of the vegetative part of a fungus?" Yes, it's about how we dove headlong into a life more in tune with our surroundings and the quality of life we wanted—in part, through mushroom cultivation. We set out to write a beginner's book on mushroom cultivation because we had to learn plenty of lessons the hard way when we were just starting to grow mushrooms and also when we started to sell them.

Other mushroom growing books have more in-depth, scientific parameters than *Mycelial Mayhem*. We've included many of these in the Resources section because they continue to serve us well as we advance in our own growing skills. You will want to gather some of these in-depth resources as you move along your path, as these are written by experts who can take your cultivation skills to the next level or share additional growing techniques. But as beginners we found many of these resources to be intimidating, as they are very technical.

Too often, books, articles, and information on a cleaner, healthier life simply ignore mushrooms. Not all species are easy to grow, but once you learn the nuances of mushrooms, they are like any other thing we cultivate. There will be banner years, there will be unforeseen issues; just like with our tomatoes and beans, mushrooms are living and breathing organisms we can embrace and integrate into our lives.

As we wrote this book, we were constantly going back and forth about what we were saying and how we were saying it. Dave came from the viewpoint that people reading *Mycelial Mayhem* will already have a basic understanding of growing things. Kristin, who ran the native plant portion of the business, wasn't sure it was that cut and dried. She would constantly remind Dave that, although she could grow some of the hardest natives to cultivate, that doesn't mean she could grow mushrooms with ease. The basic elements, though, are similar:

- Growing medium: wood chips, straw, logs for mushrooms vs. dirt for plants.
- The right amount of water is important for both plants and mushrooms—not too much but not too little.
- The right climate, weather, shading, etc.
- Learning to cultivate comes with time, love, patience, and dedication.

In our case, Dave was the mushroom laborer, the cultivator. From cutting and inoculating to monitoring and ordering spawn, he was in charge. Kristin handled the money, business operations, and marketing. Being a small organic operation and having three young children (in an economically challenging region, no less) is no easy task. Writing a book together, moving across the country, and still maintaining some semblance of sanity is even more amazing. It's the passion we have for a better world and a better future for our children that drives us.

No matter where we go across the country, we find a growing number of people who are more conscious of their surroundings. Years ago, when we mentioned that we grew mushrooms, the majority of facial expressions were puzzlement or the always-ubiquitous "Oh, the funny kind?" snicker. We notice this has changed, to "what kind?" or "which are the easiest?" or "I have _____ trees on my property, so what mushrooms can I grow on those?" People are starting to ask "Why not mush-rooms?" We saw that in our classes (the Shroom Classroom) at the talks we gave. More and more people are trying to figure out ways to have "a little fungus among us," if you will.

Most people have forgotten what they learned about the mushroom life cycle in their middle school science class. At farmers markets or shows we often hear, "My grandfather used to pick mushrooms. I wish I would have paid more attention to it or stuck with it. How did you get into mushrooms? Can you teach me?" And thus the journey into mycelial mayhem can begin. We also notice that people from small farming operations, such as cheese makers, organic veggie growers, and beekeepers are trying to figure out if or how they can incorporate mushrooms into their business models. It is heartening that more and more people are curious about mushrooms and the role they play in our world.

Mycelial Mayhem presents some basic cultivation techniques we found to be successful, sometimes after trial and error. But, *Mycelial Mayhem* is not just about mushroom-growing techniques. Bubbling up to the surface organically, like a mushroom, were the lessons we learned, reflections on what worked for us and what didn't, and how we can help you avoid our pitfalls or be better prepared, saving time, ef-fort, capital, and frustration—a little less mayhem. Also bubbling to the surface is mycelial mayhem as an integral part of a healthy, enjoyable, quality life, from physical health to landscape resilience. We share how, even mushroom growers in the earlier stages of serious cultivation can create a vibrant mushroom business or diversify their income through mushroom sales.

In Section I, *Mycelia*, we lay the foundation, considering why mush-rooms, what are mushrooms, and what do they do for us and the planet. In Section II, *The Stem*, we focus on growth, just as a stem does. The chapters of Section II cover basic mushroom cultivation, wild col-lecting, and incorporating sustainable methods for a more resilient landscape. In Section III, *The Fruit of Your Labor*, we talk about the nutritional and culinary benefits of mushrooms and how you can turn mushroom cultivation into a business or supplemental income. Section IV, *Spreading the Spores* looks at the joy of sharing and becoming a proponent of mushrooms, whether in the business or not.

We aspire, through *Mycelial Mayhem*, to entertain you with our true stories, which share the unique nuances of growing and selling mushrooms. Laugh at and learn from these stories. Hopefully, this book will be the first step on a path that is littered with mushrooms, healthy landscapes, and healthy people in your life. We welcome you to visit the book website, mycelialmayhem.com to continue the conversation and provide us your input so that we can continue to share more and more information about bringing mycelial mayhem into people's lives.

Dave's Mycelial Mayhem Journey

I didn't know how to start this book, because for me, mushrooms have always been part of my life. As far back as I can remember, I was attached to my grandfather's hip. He was from Slovakia, and he taught me everything I needed to know by kindergarten: hunting, fishing, gardening, winemaking, and mushroom picking. In my mind's eye, I can still imagine that little blonde boy scrambling through the Pennsylvania woods looking for mushrooms with his Papap. These skills carried me through my life, even after I'd gone to college. Even when I was single, I always had a garden—a postage stamp at first. I have always picked mushrooms, and for 15+ years, I've cultivated them too. For years, even with a 40+ hour per week job, I always managed to make sure that mushrooms were part of our family's diet. I wanted my children to have healthy food, because, through working in the environmental field, I became increasingly alarmed with what was offered at the grocery store.

As a family, we frequently collected wild mushrooms and would augment them with cultivated species. The more I grew in organic gardening, the more I integrated mushrooms. I was transposing the symbiotic relationships that I discovered in the woods into our garden. I was transferring the rich hummus of the Appalachian hardwood forests into the hardscrabble I was trying to garden.

In 2005, Kristin and I purchased a 3-acre property on top of the Eastern Continental Divide. It was the first property we owned, a blank slate, so the first property we could really transform the way we wanted to. At our location, we could barely get a shovel in the ground. The only way we'd be able to make money from the property was if regular old rocks somehow became valuable. With pickup truck loads of my friend's composted horse manure, a rototiller, and a lot of sweat, our garden grew. I was rotating beans every other year where my heavy feeders were: broccoli, cabbage, and corn. I started throwing used mushroom mulch into the garden, and where I threw that mulch, greener, darker leaves from healthier plants were sprouting the next year, and a few mushrooms popped up too. We were on our way to organic gardening on a grander scale and toward more sustainable ways too. The kids love to help with the planting and the eating, which is a beautiful thing! If you care what you put in your body or feed your children, why not spend therapeutic hours playing in the dirt? Our modern food system encourages us to load up on cheap, toxic, non-nutritious food, creating a society of illness and symptom treatment, rather than nutrition and prevention. So, if you want to be healthy, garden, farm, or support your local farmer!

Kristin's Mycelial Mayhem Journey

I never gave much thought to mushrooms, that is, until I met Dave. I was once out to dinner with someone who ordered chicken marsala, hold the mushrooms, which was memorable. Occasionally, I'd put some raw button mushrooms on my salad at the local salad bar, but that was about the extent of it. Now, the thought of doing so horrifies me. Raw buttons? They have no appeal to me anymore. Dave introduced me to the world of mushrooms, first with some wild-collected chicken of the woods, which I thought had to be my favorite, until I had his home-grown shiitakes. Then, those were my favorite. I had lion's mane later, and once again, had a new favorite. Who wouldn't like a mushroom that tastes like seafood and is good for your brain? Then I ate pioppino—my fourth new favorite! Thanks to Dave's passion for mushrooms and gardening, we've enjoyed years of great mushroom dishes, from mushroom lasagna to cream of black chanterelle with watercress—the best soup I've ever eaten! We both worked happily in the environmental field for years, growing our children, our garden, and our lives. When it was time for a career change, we naturally turned to sustainable farming, which, of course, focused a lot on mushrooms. Now, I have mushrooms every day, whether fresh in my diet, through supplements I buy or create, or the reishi tea I make. I feel strange if I don't have mushrooms each day because I know that I need their health benefits. I'm thankful that we get an opportunity to share our knowledge and mishaps with you and to grow mushrooms in a whole new environment with new challenges. We'll let you know how that goes, too!

MYCELIA

Mushrooms

*The muted muscles
musty thriving
musical mysteries of mycelia.*

*Fruiting bodies the tip of the iceberg
and beneath the earth, a galaxy.*

*If no one ate death
where would we get life and light?*

*Mushrooms insisting in the night
they clean the corners.*

*Masters at thriving
neon or fringed
or round and ivory as antique door knobs
shelves, flutes,
prehistoric silhouettes.*

*Hunched and giving
these neural networks of the earth
that stitch up dreams
sipping the sweetness
from stones.*

—by the Poet of the Pennsylvania Organic FarmFest,
who gifted us with her poem, but not her name.

Why Mushrooms?

In our maturation from infant to child, we all go through the "why stage." Every parent, uncle, neighbor, etc., who has been around children will know what this is. Every second of the day seems to merit a "why" question, as in "why is the sky blue?" or "why does the sun rise and set?" In some ways we remain in this stage all our life, but as we age we internalize this rapid-fire questioning. We buy books, we go online, we peruse magazines and other sources of wisdom to answer our questions. Whether it is gardening, fly fishing, painting, or playing music, we are constantly adding layers to our being, and we must go through the "why stage" as we progress through life. If you are reading this book, you might be in the "why stage" regarding mushroom cultivation. We hope to answer some of your "why's" as well as the "how's." So: Why mushrooms?

Why Not Mushrooms?

We think the question should really be "Why *not* mushrooms?" Fungi are one of the five kingdoms of life but they are easily overlooked. Without fungi, our world would be very different: giant piles of leaves, wood that never rots, no cheese, no medicine, and heaven forbid, no beer! People are getting in touch with simpler living. Conversations about organic gardening can be in-depth, educational, and passionate. Talking tomatoes can go on for hours. People can have lurid, almost

pornographic discussions about heirloom varieties—which are hardi-
est, earliest, and tastiest. But bring up mushrooms, and people talk only
about white buttons, portobellos, or maybe morels or chanterelles. End
of discussion. How is it that so many people who pour so much loving
care—their blood, sweat, and tears—into their gardens settle for industrialized mushrooms? Organic growers use a wide variety of insects,
hot pepper juice, grass clippings, and planting arrangements to achieve
their goals. Yet, no mushrooms. The fifth kingdom is too often forgotten
and is rarely incorporated into the cultivated ecosystem. We hope that
in these pages you will find ways to incorporate mushrooms into your
garden, your landscape, and into your life.

Many of us gardeners know the smell of really good compost, just
as anyone who loves nature knows the same smell from the rich humus
of the forest. In the forest, the smell is produced by the unseen, hard
work of, in large part, mushrooms. Why not put that hard work to use
to enrich your landscape and your garden? Mushrooms are amazing,
beautiful, bizarre, and, at times, ugly. But they are full-fledged members
of the circle of life that all too often are forgotten or neglected. So, why
not mushrooms?

For years, Dave grew shiitake logs as a hobby to augment his wild
mushroom gathering and enhance dishes made from our garden's
bounty. When we bought our first family house, a log cabin on top of
the Eastern Continental Divide in western Pennsylvania, Dave had a
lot of work to do. It was a blank slate, just grass. No gardens, no landscaping. He created an asparagus patch and cut down trees to make our
garden, provide firewood, and to create a supply of shiitake logs. Any
gardener will tell you that your veggies are only as good as your dirt. On
top of our mountain, rocks and dandelions ruled. If they were money,
we'd be millionaires. Dave set out on a mission: digging the rocks out of
the future garden site and making them into a water feature. He brought
many pickup truck loads of composted horse manure from our friend's
horse barn to create good garden soil. We composted, never sprayed the
property, and used grass clippings as mulch. Saplings were turned into a
woven fences around the garden or chipped to make paths through the

garden. Dave's love of mushrooms led him to the discovery of wine cap stropharia. Once we had our permanent garden site, after we incorporated wine caps, we noticed that our plants looked healthier. When we incorporated mushrooms into our compost, we derived better soil from it. So, why not mushrooms?

Somewhere along our shared human history, modern humans have turned into specialized creatures, as opposed to the "jack of all trades" our forefathers were. As we strive for a simpler and cleaner life, we must understand diversifying and therefore biodiversity. Doing away with chemicals as much as possible in our daily lives is a great start. Landscaping with natives, organic gardening, companion planting, and growing mushrooms are all part of a healthy lifestyle. Biodiversity helps to create resilience in systems. Without the fifth kingdom, we cannot achieve great biodiversity and therefore cannot achieve great resilience in our systems (see Chapter 5 for more information on resilience). Plus, biodiversity enriches our lives. If we have a variety of good food to eat, we feel more alive. If we can go on hikes and see a variety of plants and animals, we feel more alive.

We grow and sell vegetables at farmers markets. Why was our kale favored over all other vendors' at farmers markets? Why did people reserve our kale via Facebook messages days before the market? We firmly believe it was the magic of wine caps thriving in our soil. Perhaps they were "sipping the sweetness from stones," as our anonymous poet so eloquently wrote. Without the collaboration between these two different species, we wouldn't have had the sweetest kale and tastiest wine caps around. So, why not mushrooms?

Mushrooms Are Part of Our Fabric

In September 1991, two German tourists hiking in the Alps found "Otzi the Iceman," a mummy frozen around 3200 BC. What does this have to do with mushrooms? Well Otzi, who died about 5,300 years ago, was carrying *Fomes fomentarius* or amadou, a mushroom known as the tinder fungus, which can be used to start fires. The Iceman was using mushrooms in his journey crossing the Alps! (Dave knows the amadou because he uses it to dry his fishing flies, because it extracts water excellently.) So, mankind has utilized mushrooms since the dawn of humanity.

Further back in the human history book, 7,000–9,000 years ago, in the Sahara Desert, rock paintings depict mushrooms and gods covered in mushrooms, clearly illustrating the long-running connection between mankind and mushrooms. In some of these historical cases, mushrooms (specifically, the hallucinogenic kind) were used in rituals and rites of passage. From the female shamans of Siberia with their mushroom headwear to Africa and Mesoamerica, mushrooms have played a significant role in human history. Mushrooms containing hallucinogenic psilocybin and *Amanita muscaria* were consumed during rituals. In Chinese history, lingzhi (*Ganoderma lucidum*), what we call reishi, is mentioned in writings as early as the fourth century BC. Maybe Lucy's ancestors came out of the tree to try *Termitomyces robustus*, a mushroom found in Africa that grows on termite mounds, and is still a favored edible there. Mushrooms are woven into human history.

It seems that in certain cultures mushrooms are just part of the fabric and nothing out of the ordinary, but in others they are nearly nonexistent. We have noticed that people of Eastern Europe, Asia, Italy, France, and some regions of Germany utilize mushrooms as a large part of their diet. As we got more into cultivating and selling mushrooms, we recognized how Chinese and Japanese cultures really understand mushrooms—on so many levels—and are light years ahead of Western culture. Dave personally noticed a difference in cultures when he married Kristin, whose family didn't regularly consume mushrooms. Dave, being of Slovak, German, and Serbian ancestry, had mushrooms on the table as far back as he can remember. It wasn't until he was older that he realized that his mother, a pure Slovak, was the source of the mushroom love. Since Kristin met Dave, she has become the medicinal mushroom queen, constantly studying their effects and reading about any new discoveries about them. She is intrigued by the medicinal aspects of mushrooms and plants as natural remedies, even using mushroom facial cream to keep her skin healthy and as wrinkle-free as possible. Dave thinks more with his belly. He loves to cook with them, as they bring a rich layer to so many dishes, with their different flavors and textures.

Mushrooms for the Future

With the myriad of health and environmental challenges of today, mushrooms should be an integral part of our future. You ask, "Mushrooms…can solve problems?" Yes, mushrooms. Plus, we're just missing out if we don't incorporate mushrooms. We're missing the "umami" of life, the pleasant, savory aspects that make our lives better, more enjoyable.

Did you know these fungi facts?

- A honey fungus in Oregon is believed to be the largest living organism in the world, spanning over two miles in the Blue Mountains.
- Some mushrooms have demonstrated anti-tumor activity. Yes, you read that right! Some even show the potential to shrink tumors.
- Mushrooms serve over 100 health-promoting functions, contributing to long, healthy, happy lives.
- "Umami"—that pleasant, savory, even meaty taste of mushrooms—can reduce your need for meat and salt, therefore improving your health.
- Mushrooms alkalize the human body, providing a healthy, protected internal system of slightly alkaline pH.
- People in the United States are consuming four times the mushrooms today than they were in the late 1960s.
- Huge profit potential exists in mushroom purveying, from selling fresh mushrooms to developing new, innovative mushroom products. The sky's the limit.
- Mushrooms in your garden can enrich your soil, making all of your veggies better tasting and more nutritious.
- Mushrooms work to decompose organic matter in nature, therefore cleaning up the environment for free.
- Mushrooms have the potential to restore landscapes by soaking up pollutants; the process is called *mycorestoration* or *mycoremediation*.

- Mushroom mycelium and agricultural waste are being combined into packaging material that can be molded into many forms, perform like foam, and can be composted.
- Mushroom mycelium gives off CO_2, and plants need CO_2; so, you can augment your greenhouse growth with mushrooms. This is an example of *mutualism* at work.
- Fungi and the many products made from them can help us close loops—moving from linear, wasteful systems to cycles, as nature intended.
- Mushroom cultivation can help us to play the diversity of roles we're supposed to—getting us back to the basics.
- Fungi have the potential to outcompete undesirable kinds of fungal infections, both within the human body and in ecosystems, therefore serving yet another health function and contributing to resilient, organic ecosystems.

These are only some of the ways fungi can play an integral role in a successful future for mankind. Can you think of other ways? Get creative. The sky's the limit.

On the Mushroom Path

We can learn from mushrooms. "Hunched and giving," our anonymous poet writes. Mushrooms are not just food. Mushrooms are not just decomposers. We can learn from them. How can their "galaxy," the "neural networks of the earth" teach us about human networks and collaboration? How can mycorrhizal fungi relationships in nature teach us about mutualism in society? What about sporulation? Mushrooms share their genetic material and promote a new generation this way. What could this teach us about sharing, looking toward the future, and creating a new generation of stewards?

Although mushrooms have been a part of human history as far back as when we dropped out of the trees, we don't yet understand their ways. From the elusive cultivation of morels to the symbiosis between the mushroom and the termite, mushrooms are still confounding mycologists—adding to their mystique. So many questions about mushrooms still need to be asked and answered.

In this book, we offer ways to include mushrooms in your life, a starting point along your journey. Once on this path and you begin to have conversations with like-minded people, you will realize and learn what works for you and what potential your mushrooms might have. The journey along the mushroom trail is as varied as mushrooms themselves. You might end up on a path that leads to becoming a wild purveyor who walks and stalks in the woods; or you might find yourself wanting to become a full-blown mushroom cultivator working on the mycorestoration solution. There is something for everyone. Each individual can customize their mushrooming to their needs and wants.

We could just ignore mushrooms, but where is the fun in that? Harvesting and cooking with them is a culinary art form, and selling your extras could help make or break a farmers market for the small organic farmer. The trailhead, a starting point, *Mycelial Mayhem*, we hope, conveys what you need: information and inspiration as you embark on your journey into the fifth kingdom!

Mushroom Basics

Fungi: A Kingdom of Its Own

As we go about our lives here on Earth, we seldom think about the natural world and what it provides us, let alone about the significant value and role that mushrooms and other fungi play in providing us with the clean, healthy ecosystems from which we derive our food, shelter, clothing, and other life necessities. Scientists have developed a classification system for all living things. Depending on which classification system you subscribe to, there are five to eight kingdoms of life on Earth (Table 2.1). Although scientists may disagree on the number of divisions, they all feature Fungi as a major division. Kingdoms are the second highest classification of life (beneath domains) and describe some of the most basic divisions, for example, animals vs. plants.

The five major kingdoms (Whittaker, 1969) are:

Plantae: includes eukaryotic (cells contain a nucleus of genetic material and other organelles, all enclosed in membranes) organisms that are multicellular and are usually characterized by photosynthesis and sexual reproduction, and is always its own kingdom within all classifications to date.

Animalia: includes eukaryotic organisms that rely in some way on other organisms for nourishment, and is always its own kingdom within all classifications to date.

Fungi: includes eukaryotic, non-photosynthetic organisms distinguished by unique cellular wall structures, primarily the presence of

TABLE 2.1. The Kingdoms

Five Kingdoms	Six Kingdoms	Seven Kingdoms	Eight Kingdoms
Monera	Eubacteria	Eubacteria	Eubacteria
	Archaebacteria	Archaebacteria	Archaebacteria
Protista	Protista	Protozoa	Archezoa
			Protozoa
		Chromista	Chromista
Plantae	Plantae	Plantae	Plantae
Fungi	*Fungi*	*Fungi*	*Fungi*
Animalia	Animalia	Animalia	Animalia

chitin rather than cellulose, and is comprised of mushrooms, yeasts, and molds, and is always its own kingdom within all classifications to date.

Monera: includes all bacteria, which are single-celled, prokaryotic (cells do not have a distinguished nuclear membrane) organisms, and is sometimes broken into two kingdoms, usually Eubacteria and Archaebacteria (Woese, 1977).

Protista: a catch-all (not plants, animals, fungi, or bacteria), diverse kingdom of eukaryotic organisms with a simple cellular structure; it has been divided into two or three separate kingdoms by some biologists (Cavalier-Smith, 1993 and 1998).

Fungus Fact: The largest single organism in the world is thought to be a honey fungus that spans 2.4 miles (3.8 km) within the Blue Mountains of Oregon.

Mushrooms are a small group of species that exist within the Fungi Kingdom. Although the term "mushroom" is not a biological classification at any level, it is used loosely to describe fungi that have a visible fruiting body, which includes a stem and a cap with either gills or pores on the underside that produce reproductive spores. Within ecosystems, the primary role of fungi is decomposition, but they serve a variety of other functions as well. We cultivate mushrooms that are different species (boletes, stropharia), but can also be sub-species or varieties of the same species that go by different variety names.

Many species and a wide variety of mushrooms exist. Let us start with the ones you're probably most familiar with: white button, crimini, and portobello. These mushrooms are actually the same species, *Agaricus bisporus*. In most grocery stores, your fresh mushroom selection is limited to *A. bisporus* because its production has been industrialized. The large tan version of this mushroom was a waste product at one time. It wasn't until the 1980s and early 1990s that mushroom growers started to market this variety as the now-ubiquitous portobello. The two of us set out on our mushroom-growing path because we felt *Agaricus* mushrooms are not that tasty and are often not as nutritious as other mushrooms. Plus, they must be grown on manure compost, and they can contain hydrazines, which are cancer-causing chemicals (FC&A Publishing, 2004). We now wild-collect mushrooms and grow gourmet and medicinal mushrooms that feed on plant cellulose, rather than manure compost. This is foundational to our mushroom journey and, therefore, the contents of *Mycelial Mayhem*.

Mushroom Structure and Life Cycle

The word "mushroom" most often refers to the *fruiting body*, the visible part of the organism, with the *stipe*, which is like its stem, and the *pileus*, or cap, which sprouts first as a small *pin* (technically called the *primordium*), which looks like a tiny bump bubbling up from the parent *mycelium*.

But, let's back up to the beginning of the life cycle, the germination of a mushroom's *spores*, which are like a mushroom's seeds. When the spores are released from a fungi's fruiting body, usually from underneath the cap, they germinate, and start to form threads of cells, called *hyphae*. Hyphae continue to

FIGURE 2.1. Oyster mushrooms primordia (pins) starting to form fruiting bodies (stipe and pileus).

grow and branch out, forming mycelium. Mycelium is a network of cells that spreads like a plant's roots within the organism's growth medium, usually soil, manure or various types of plant material such as straw or woody substrates. From the mycelium, the pins arise from cells that gather and push outward and upward. These form the fruiting bodies, usually comprised of the stipe and the cap. Spores mature within the fruiting body, usually beneath the cap, and are released when ready.

When we exhibit at shows or teach classes, Dave usually wears a black t-shirt that has the mushroom life cycle on it. All he has to do is point at the shirt when he wants to explain the process of a mushroom's life cycle.

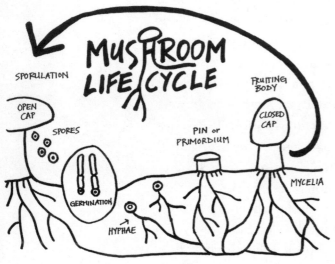

FIGURE 2.2. Lifecycle of fungi.

Relevance of the Life Cycle to Cultivation and Harvest

Here, we'll cover the basics of mushroom growth and their successful cultivation and harvest; we'll expand on this information in Chapter 3. A good understanding of mushroom properties and life cycle will help you have cultivation success, capitalize in various ways on these features during cultivation, and think more broadly of the inherent wisdom of the mushroom life cycle, which can (somewhat surprisingly) benefit your business perspectives as well. As Paul Stamets and J.S. Chilton explain in *The Mushroom Cultivator* (1983), you, the cultivator and harvester, interact with all stages of the mushroom's life cycle and therefore need to know how to facilitate the cycle, usually with the goal of producing as many mushrooms as possible, as frequently as possible. Unfortunately, successfully creating a healthy growth environment for say, oyster mushroom mycelia, can also create an excellent home for green mold and other invaders. So you will need to learn how to tip the scales in favor of the mushroom you are trying to grow. In the case of the oyster, you do this by pasteurizing your straw, keeping everything as clean as possible, and not getting too much water into the production bags—because oysters require less water for their life cycle than green mold.

If you provide just the right environment for your mushrooms, you will literally see the fruits of your labors. The first thing you'll see will be healthy *mycelia* colonizing your growing medium. The figure here shows an Italian oyster "production bag" completely populated with oyster mycelium, which appear in the photo as solid white. Understanding what a mushroom looks like at its various stages is important. You need to know when it is releasing its spores so you can collect them at the right time, and you'll want to recognize stages to help with the timing of your harvest. For example, some mushrooms, like wine cap stropharia, become bitter-tasting when their caps open and start to produce spores.

Cultivating mushrooms is as exciting and fun as gardening or any other effort in which we shepherd our own food creation. We are not so much the masters of the processes but the caretakers of these ageless cycles of life. Much like picking tomatoes at the height of perfection, you will want to know the perfect time to pick your mushrooms. You want to pick for optimal taste and shelf life. Every mushroom and every strain is a little different. For example, hericium (lion's mane) is very different from shiitake in its structure. One of the most often wild-collected mushrooms is *Coprinus comatus*, known as "shaggy mane," "lawyers wig," or "inky caps." These edible mushrooms will go from a scaly whitish beauty to an oily black mess of spores in a couple of days. For optimal taste and before they turn oily, "ya gotta get her before she lifts

FIGURE 2.3. Mycelium throughout the bag produces plenty of Italian oysters.

her skirt," in other words, before the veil is broken and the spores start to mature.

With our cultivated mushrooms, these same processes play out, so as a cultivator, you will need to know when to pick, when to compost, and when to allow the full process to play out. Often, we allow wine cap stropharia to play out so that it will sporulate, continuing its life cycle and fostering future grow beds. Because of the fast growth of the fruiting body, though, we sometimes accidentally allowed it to play out. But that's ok, because we know this will strengthen the beds and augment our garden soil. When eating or picking to sell wine caps, we harvest the fruiting body when the gills are white and the caps were still somewhat closed and burgundy in color. This is important because taste is optimal before sporulation starts. This demonstrates the importance of not only knowing a mushroom's life cycle, but also what each of your mushrooms looks like at various stages in the cycle.

FIGURE 2.4. Wine caps harvested at the right time.

The fruiting body is what we sell, and therefore what we are known for. The average customer is not interested in the mycelial structure or timing; they just want to see fresh and beautiful mushrooms. So, let's start the discussion with varieties that have the "classic" look of the mushroom fruiting body: shiitake, pioppino, and wine cap stropharia.

Where the mycelium breaks from the ground, log, or substrate, is called the base. Next is the stipe or stalk. A stipe can be solid or hollow (as is the case with morels). Some call this the stem. If there is a "netting" look on the stipe, this is called *reticulum*. (Many

FIGURE 2.5. Wine caps fully open, left to spread spores.

FIGURE 2.6. Shiitake grow block, showing various stages of growth and parts.

boletes have this on their stipe.) If there is a ring around the stipe, of tattered or papery appearance, this is the *annulus* and is where the cap's veil was attached to the stipe. You will see an annulus on stropharia, shiitake, and pioppino, too. The "cap" of the mushroom is known as the pileus. On the underside of the pileus, the *hymenium* exists in the forms of *gills*, technically called *lamellae*, *tubes*, or *pores*. (Shiitake, stropharia, pioppino, oyster, and nameko all have gills, while boletes have pores.) For the best harvest of your cultivated mushrooms, you want to collect

FIGURE 2.7. Wine cap annulus.

FIGURE 2.8. Shiitake's hymenium, in the form of gills (lamellae).

your mushrooms right before or as soon as possible after the cap's veil breaks from the stipe. This happens prior to sporulation, which compromises most mushrooms' taste. Once the veil breaks, the caps tend to flatten out to facilitate spore release. Harvesting at this time also gives your mushroom a longer shelf life. However, we once had a customer who liked shiitake fully flattened out!

But mushrooms are all different. With oysters, you do not have the luxury of looking to see the veil breaking. You have to watch them and learn when to pick them. Oyster caps will also begin to flatten out as the mushrooms begin to release their spores. The best way to know when to pick oysters is to watch the caps. The mushroom caps will drop a "dust"—the spores—on their neighbor below. This means that sporulation has begun. Once you start to see any spores on any of the caps in an oyster cluster, pick them! It will only be a matter of hours or a day or two before every cap is covered in spores and the taste will be compromised. Oysters are notorious producers of copious spores.

With hericium (lion's mane), you will watch the top of the mushroom more than anything else. A small ball will push out of your wood or from the substrate, and soon it will grow "teeth." Once the top of hericium begins to turn brown or becomes discolored, it is time to harvest the mushroom.

Maitake will "frond out," meaning it will have plates or leaves that will be slightly upright at first but will flatten out. We always try to pick our cultivated maitake just when the fronds begin the flattening process. In the wild, we grab them any way we can, or we watch and wait for the flattening process. If we see a maitake we want to pick, we mark it and then watch the weather. If a drencher rain storm is coming, we pick it. If very hot weather or severe uptick in temperature is on the horizon, we pick the mushroom. If a hard freeze or snow event is coming, we grab it. Otherwise, we allow it to run its course to the flattening stage.

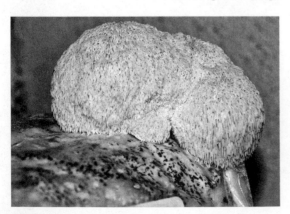

FIGURE 2.9. Lion's mane ball ready for harvest.

FIGURE 2.10. Maitake fruiting, primordia (*left*) and fruiting body formation (*right*).

FIGURE 2.11. Maitake fruits, fully "fronded out" and ready to eat.

Facilitating Growth and Reproduction

Most beginning mushroom growers purchase spawn from a supplier. Spawn is a mixture of materials that contains all of the necessary ingredients to facilitate mycelial growth and, eventually, mushrooms. Ingredients include an initial growth medium or food for mycelia, usually a grain. Mycelia has usually already started to spread within the grain upon purchase of spawn. Once received, the grower distributes the spawn within more growth media, such as wood chips, straw, or wood. Knowing what a mushroom's mycelia looks like will help you recognize good mycelial growth as well as growth that deviates from that, which can indicate an invasion of some other fungus that is enjoying your growth medium and may outcompete your mushroom.

If you decide to move into the production of mycelia from spores, with all of the laboratory equipment and knowledge necessary, you will need to know how to recognize sporulation so that you can collect spores. You'll then facilitate their growth within a sterilized medium, usually agar in a Petri dish. To recognize sporulation with the naked eye, look for the cap of the mushroom to start opening or spreading out laterally, exposing the gills or pores underneath so that the spores contained within can be more easily released.

Much more knowledge than this is needed to successfully grow mushrooms from spore to cap, so see Chapter 3 for more information and the technical resources you will need to succeed. Still, recognizing when a mushroom cap is in its pre-sporulation phase is important for any grower, as this is the best time to harvest for freshness and optimal taste. The signs vary by mushroom, and some produce spores a lot faster than others, so learning what the mushroom looks like in various phases is important.

Beyond producing mushrooms, knowing about the structures and life cycle can help you on other fronts as well. For example, wine cap stropharia can serve as a great companion within a vegetable garden. The mycelial mat that forms within the straw on the ground can hold important moisture in the soil, providing water for adjacent plants. The fruited wine cap bodies that are not harvested in time to eat (which can be numerous since they are fast growing) can provide much-needed nutrients to your soil as they quickly degrade on the ground.

A mushroom's mycelial mat can serve a variety of purposes. Mycelium can be used as a dietary supplement because the structural and chemical properties of mycelium can mirror that of the fruiting body. Many times, it is not the fruiting bodies, but the mycelia that are the raw materials of mushroom supplements because mass-producing an ongoing mycelium source can be more reliable than the short-lived fruiting body.

Fungi in Our Ecosystems

In ecosystems, everything is connected, interdependent. So, it is logical that the species that make up one of the five kingdoms of life play many important roles, filling *niches*, as they are known in ecology. Most people know that mushrooms are the master decomposers in nature, filling a large portion of this huge niche so that we don't have too much "stuff" build up in nature, like fallen leaves, trees, and other organic matter. Mushrooms feed on this organic matter, cycling nutrients back into the ground upon their demise. But, as if they don't do enough as decomposers, fungi also serve other roles in nature. An example of mutualism (mutually beneficial relationships), mycorrhizal fungi facilitate plant growth by providing essential nutrients to a plant's roots, while receiving sugars from the plant. According to Gary Lincoff (2015), fungi also form other mutualistic relationships with plants, such as by occupying plant tissues (becoming *endophytes*) and potentially contributing to their functions. They also contribute to soil ecology and provide the "network" of underground mycelia so important in the exchange of resources. It's the underground Internet, so to speak. So, fungi contribute both to the healthy existence of plants and the breakdown of excess plant material.

Fungi can also cause system disturbances, as seen with the ergot fungus; commonly known as St. Anthony's fire, it caused widespread illness and rye crop devastation in Europe during the Middle Ages. Another example is Dutch elm disease, an infection caused by sac fungi (transported by the elm bark beetle) that has all but wiped out the American elm tree from North America. It's important to note that this is a case of non-native invasion; the disease was brought from Asia to North America and Europe, where elm trees did not have resistance to the disease.

Though we still have much to learn about all these interactions, we already know enough to determine that fungi are vital to the existence of life on Earth, despite the examples of disturbance. Through the years, much more attention has been given to the negatives of fungi in our world. *Mycelial Mayhem* focuses on the many positive roles fungi play and how to incorporate them into our lives.

Mushroom Life Cycle and Ecology as Teacher

This book is organized into sections paralleling the mushroom life cycle because it's a great analogy for the journey of mushroom cultivation. The mushroom life cycle can be a process that reminds you of the important stages in your new mushroom business. The genesis of your business requires the basic ingredients of desire and "seed" resources to begin, just as the mushroom requires spores coming together to form hyphae and then mycelia. Once you have the basics germinating

together, you need to prepare for a fruitful business by learning, growing, and branching out your foundation, the hyphae. Developing a game
plan is required for a fruitful business, as you need all of the energy,
food, and resources you can muster. This is your mycelial growth stage.
Once you have a solid foundation, you can produce great results, the
fruits of your labors. But, you must always keep an eye on your foundation for challenges and obstacles that might set the stage for competition to take over. Sporulation signifies sharing or spreading of your
mushrooms and your fungi passion and knowledge, whether you're
simply sharing with friends and family or growing a successful mushroom business.

There are things to learn from fungi's roles in ecosystems. We know
they form mutually beneficial relationships, ones that benefit both participating parties. Can you participate in similar mutualism? Can you
barter your mushrooms for organic straw? Can you form two-way partnerships with other organic farming businesses, such as working with
a beekeeper to co-package and sell reishi tea kits that include reishi,
honey, and other ingredients? In highly evolved natural ecosystems,
collaboration happens all of the time, more than originally thought,
perhaps even more so than competition. Collaboration is the way of
the future. We are wise if we look to nature—including the world of
fungi—for inspiration and applications to human society.

THE STEM

Growing Mushrooms

So, you are interested in learning how to grow mushrooms? Do you want to grow them as a hobby and have a fresh supply for your family and friends? Would you like to add income to your farm? Whatever your reasons, you need to know what is involved and where to start. You'll first need to do an assessment of what you have readily available in terms of time and money, and what you can put into your mushroom growing endeavor. Take growing shiitake, for example. It takes quite a while (from six months to over a year) to transform a fresh-cut log into a harvest of fresh mushrooms. So, you have to take into account what you want for yourself or your business. When we started to seriously market our mushrooms, Dave had already been a mushroom hobbyist for close to 20 years, so he had a working knowledge of the different varieties and what it takes to grow them on a regular basis. This helped us tremendously!

The Assessment

The first thing to do is figure out what you have to work with. Following are the kinds of questions you'll need to answer:

- How many acres do you have?
- What tree species do you have?
- What water source(s) are available?

FIGURE 3.1. Fall 2013 Shroom Classroom students.

- If your property is limited, or is a small lot dominated by lawn, what would you have to do to create a shady area you can water?
- What substrate do you have ready and affordable access to, and which mushrooms match up best with that growing medium?
- Do you have a shady place to house stacks of shiitake logs?
- If you don't have trees, do you have a source for affordable straw, cotton seed hulls, scrap logs, or wood chips, such as from an ag center?
- How much money are you willing to spend to get rolling?
- Do you have friends or family who are willing to help with the labor for a share of the harvest?
- If you're planning on selling mushrooms, can you research who might want the mushrooms you grow and which ones they are likely to want most? (See chapter 8 for additional business planning.)

Take some time to consider these questions. Even if you decide you're in the position to just go ahead and order spawn, call a lumberjack to cut your logs, and have a farmer drop off straw bales, answering these questions will give you a good start down the road of mushroom-growing mayhem. And if you're not in such a position, you'll save yourself time, trouble, and money if you think things out thoroughly.

For us, a variety of factors led us to concentrate mostly on shiitake and wine caps at the beginning of our business adventure. During our assessment, we found that what we needed were species with quick turnaround times from spawn to fruiting body. Plus we wanted species that would be easy to work with. And, because not everyone is a mushroom fanatic, we wanted to stay away from the more exotic species that might scare off potential customers. We realized that shiitakes were the most sought-after of the species that we already grew, so it made sense to increase our shiitake production. We started with approximately 30 inoculated logs for our first year at the markets. (Thirty logs may sound like a lot, but our runs sold out quickly!) We already had some wine cap stropharia beds on the property, so we concentrated on upping production of those as well. We also wanted to offer wild-gathered mushrooms, even knowing from the start that these are very dependent on Mother Nature, who can reward you greatly or deny you just as easily.

We realized that if we were to do this as a business, we would need a constant, readily available supply of growing medium, particularly wood, wood chips, and straw. We determined that we could meet part of our need for hardwood logs and wood chips from our own property and the rest by partnering with our logger friend, specifically by scavenging what he called "garbage" wood from his job sites, which consisted of the unmarketable treetop logs. We used some of the larger-diameter material as shiitake (and other) logs. We invested in a wood chipper so that we could also use smaller material for wine cap beds. For straw, we scoured local advertisements for the best deal and soon found a few affordable sources nearby.

One of our biggest decisions at the start was *where* to grow. It's funny how things work out, sometimes. We are conscious of our environment and the impacts we have upon the Earth. So, when we purchased our house, we decided early on to replace its coal-burning stove with a cleaner system; we switched to a high-efficiency electric hot water heater. Beyond the environmental factor, we hated dealing with the ash and the smell, and the dust was destroying Dave's fly-tying materials. Furthermore, after close to two decades of volunteer work cleaning up streams near abandoned coal mines, the coal stove simply did not jive with who we are. Once we ripped that beast out of the basement, we had a big open space, and we had a coal bin. After power washings, scrubbings, bleaching, drying, and many coats of paint, we had a whiz-bang grow room. Once we added a humidifier, a small heater, hygrometer, and fan to keep the parameters of temperature and humidity constant, we had a clean indoor grow room! The initial financial investment was minimal, and what was once a nasty coal bin would now be putting out mushrooms.

We made a conscious decision to grow indoors for economic reasons. But we suggest that most folks start their adventure of mushroom cultivation outdoors. However, if you are serious about full-time production, especially if you don't live in a climate that is very friendly to outdoor growth, you will need to explore indoor cultivation. If you are looking to grow specialized mushrooms (hericium, maitake, reishi), indoor cultivation is a great way to learn these mushrooms' growing parameters.

Outdoor Cultivation

Outdoor cultivation is the easiest introduction to mushroom cultivation. All you really need is some space, a substrate, and a way to keep your mushrooms shaded and moist. Dave started years ago inoculating oak logs to grow shiitake. In winter, when he would cut trees for the next year's firewood, he'd select nice, straight pieces to use as inoculation logs. A warmer-than-usual March weekend would call him out to sit on a wheel of oak and inoculate logs. For Dave, it was a way to cure cabin fever when it was still too cold to travel to a favorite trout stream and the snow was too sloppy to go cross-country skiing. Dave started with shiitake, then moved to oyster, maitake and finally wine cap stropharia. Dave's never been one to take the easy road, so don't follow him here. Novices should consider revising this order to something more like: wine cap stropharia, oyster or shiitake, and then maitake, primarily due to their degrees of growing difficulty.

Considering Dave's fascination with the ways of "the old country" and a simpler life, he should have known to start with the stropharia. He treasures an old VHS videotape of his Papap and Nana when they visited Slovakia. In one sequence, family members, women with babushkas, and the white shepherd dogs of the Tatras Mountains were trekking up into the alpine fields with baskets and walking sticks. Nana was smiling from ear to ear and holding a basket full of mushrooms. What Dave didn't recognize was that most of the burgundy-colored mushrooms in that basket were stropharia! In many areas of Eastern Europe, where straw, corn, and other crops were once grown, stropharia has taken over fallow fields. As long as the substrate is refreshed from time to time with woody debris or straw, this "garden king" will continue to grow. Because of its quick growth rate, simplicity, and great return on investment, most beginning hobbyists and entrepreneurs should consider starting with this mushroom. Plus, we assume that many of you already have a garden, which is where this species can thrive on hardwood chips or straw. Let's introduce this beauty and then go over the two primary methods we've used to grow this great mushroom. (We discuss stropharia in greater detail in the companion planting section of Chapter 5, including what to plant it with, its beneficial effects on veggies and soil, and more information on the best methods for growing it.)

You could have success growing only shiitakes or maybe two varieties of mushrooms, but we will show you how to grow a substantial variety of mushrooms. Starting with wine caps, we'll introduce you to the mushrooms we've cultivated and found to be popular, the steps to grow them, and lessons we've learned along the way. Because this is a beginner's resource, most of these species and varieties are easy to moderately difficult to grow. For each species, we indicate its cultivation difficulty, both indoors and outdoors, as: easy, moderately easy, moderately difficult, or difficult. For additional and more advanced species, as well as different cultivation techniques, we recommend that you move onto other, more intermediate resources, such as *Organic Mushroom*

Farming and Mycoremediation by Tradd Cotter, and *The Mushroom Cultivator* by Paul Stamets and J. S. Chilton. More cultivation books are included in the Resources section.

Wine Caps (Stropharia rugoso annulata)

Outdoors: Easy

Indoors: Moderately Easy

You will hear them referred to as the wine cap, garden king, king stropharia, burgundy, and holy shit…is that thing for real? When allowed to grow to its full maturity, it is a 3–4 pound specimen, a foot or more in diameter, with grey gills and thick stalk, that will wow you and your farmers market customers. (Note: A wine cap this large is no longer tasty, so you could just use it for display purposes.) Stropharia is a hardy mushroom that travels well, has the traditional mushroom appearance, and offers a mild taste.

There are many ways to grow this beauty. Here, we offer two of our favorites.

Method I: Wine Cap Lasagna

Although wine cap lasagna from the oven would be great too, here we offer a *gardening* recipe. For best results, you should first look long and hard at your garden. Stropharia likes a rich substrate of woody debris, a shaded locale and moisture. In our garden, we found that spot in our vegetable beds. The most in-demand product we were growing was kale (red Russian to be exact). The shade provided by the plants was just what the stropharia wanted, so we grew it in between our eight 28-foot-long rows of kale. To establish the colony quickly and have mushrooms in the fall, we installed a "lasagna" of substrate and stropharia spawn in the early spring when the kale was planted.

Instructions

1 Start by mulching a thick 1"–2"-deep layer of hardwood mulch. (We made our own mulch. As we cleared a spot for the chicken yard and the greenhouse, we cut down the trees and saplings and

Materials
- Wine cap spawn (a 5.5 lb bag of spawn will inoculate a 20' row using this method)
- Untreated straw
- Untreated hardwood mulch

fed them into the chipper/shredder. We always tried to separate out the black cherry, which we had other uses for, so we'd be left with chips of oak, maple, and beech. If you don't have the capability to generate your own wood chips, store-bought mulch works just fine too. Or you might look around for local sources of hardwood chips. Just remember: you want natural, uncolored, untreated mulch.)

2 After you lay down a row of hardwood mulch, add a layer of straw.

3 Sprinkle your spawn on this straw layer.

4 Add another layer of hardwood chips, another layer of straw and spawn, and top off with hardwood chips.

5 Water generously, keeping it constantly moist after the first thorough wetting. We used drip hoses down the rows on fresh beds to allow a slow drip. We would let them drip before sunset for approximately an hour.

FIGURE 3.2. Wine caps (*Stropharia rugosa annulata*).

If you till your garden the following spring, do not be alarmed that you will ruin your "lasagna." Just remember to add a new layer of mulch. Once established in your garden, stropharia will continue for years, as long as you add fresh substrate each year.

Dave developed this method because the spawn can run ("run" refers to mycelial growth, which includes consumption of substrate for energy in preparation for fruiting) fairly quickly through the easily "digestible" straw, depleting its food source too soon. But you can grow stropharia on just wood or straw. Growing on wood chips takes longer for the spawn run, but produces bigger, thicker mushrooms for a longer period of time. Straw-grown stropharia will run and fruit quickly,

FIGURE 3.3. Wine caps exposed to the sun with open veil.

but the mushrooms aren't as robust, and the fruitings will not last as long. By employing the "lasagna" method, you'll get the best of both worlds. If you establish your bed in early spring, you should have mushrooms in the fall—if Mother Nature blesses you with the right weather and conditions. If you don't get fruiting the first fall, be prepared for them in the spring.

Stropharia only get their rich burgundy color when they are shaded. When they are exposed to sunlight, they will take on a buckskin or tawny look. In April or May, before our garden has taken off, the wine caps will often pop up with burgundy/red caps that soon fade to the buckskin color. Stropharia are very tasty mushrooms. We prefer them in the button stage, which sometimes means we have to act fast. They will grow to maturity quickly, and once the veil breaks open (cap separates from stem and spreads out flat), the spores will start to mature, rapidly turning to a grey color. Once they hit this stage, we find them bitter. When conditions are right, you can walk out in the morning, gather a bag of buttons, and then collect more again in the afternoon. You'll also see the ones you missed. Their caps will be lighter in color and flattened out, with the gills turning grey. In the fridge, protected in a brown paper bag, your wine caps will last for a while, probably close to a week.

Method II: Mothership

The second method is the "mothership," or the pie method, which was the first cultivation method we used. (The spawn used in the lasagna method was produced using the mothership/pie method.) We developed many stropharia beds around our property, but most came out of one, original bed, hence the name "mothership."

The mothership method is for the patient type, as only one bag of spawn is used to create a slow-growing source for future patches. If stropharia is going to be your first mushroom, take some time to plan out your area or yard for cultivation. You want to readily be able to provide water (ideally, with rain barrels).

We already had an area next to the shed where we had pallets for shiitake logs. We had moved them from the woods beyond the garden to the backyard where we had easy access to a hose, and where they were closer. Dave was working in this area every day, and most of our mushroom production came from the backyard. Beds and experiments popped up wherever the garden hose could reach. For the mothership, all he had to do was walk about ten yards into the woods from the shiitake pallets and nearby oyster boxes to a witch hazel tree that provided a 360-degree low-hanging shaded area most of the year. If you can find a tree or shrub you can get "up and under" that provides shade, it will be a great spot.

Instructions

1. Lay out cardboard around the base of the tree, approximately 3' out from the trunk. The cardboard helps to prevent every competing mushroom on the hardwood forest floor from finding the spot to be perfect, too. (When he started, Dave did not realize how aggressive stropharia are, but this is still a good precaution.)
2. Cover the cardboard with a 3"–4"-thick layer of hardwood mulch.
3. Mix the bag of spawn in with the hardwood mulch.
4. Cover with cardboard. (Dave says: "If your wife's picky about looks, like mine is, you can cover the cardboard with a light layer of mulch or leaves.")
5. Place small stones or firewood chunks to hold down the cardboard in windy conditions.
6. Wet thoroughly!
7. When the cardboard gets dry on top, wet it down again.

Materials
- Wine cap spawn—2.5 lb. bag
- Untreated hardwood mulch
- Plain cardboard

Established in spring, a mothership should have mushrooms by fall. But this bed isn't for production. Once established, this bed can be used to create new beds. You will know it is established when you peek under the cardboard and see a thick bed of mycelium (the white root-like structures) and when big red caps pop out from between the edges of the cardboard layers. That is always a tell-tale sign your bed is established…fruiting is a dead giveaway!

Once this bed is established and you want to create other beds (using the lasagna method, perhaps), simply cut out a pie wedge much like you would a pizza, including the cardboard layers. Your bottom layer of cardboard is going to be more in a pulp state than a solid structure you can lift out. So, slide the "slice" out onto new cardboard and transport it to where you want to establish a new bed, such as where you want to add another row of mulch somewhere in your garden, like around a couple zucchini plants or tomatoes. Over a period of a few years, our entire garden became a wine cap bed via this method. By adding this established mycelium to a straw or wood-chipped area, you can establish another bed quickly. Note that your new bed will probably not fruit right away. You will likely see fruitings the following spring or fall, depending on when you established the new bed. After you have established your new bed, don't neglect momma! Insert some new cardboard into the bottom layer of the mothership, and add your substrate of wood chips or straw, and cover with more cardboard. It was through this method that Dave developed the wood chip/straw/lasagna method, which allows the fast run and hardier mushrooms and mycelia than just the straw. By first establishing a mothership bed you can, over a number of years, create numerous beds in shaded areas of your property and within your gardens. Soon you will have production in quantities large enough to sell at markets, restaurants, etc.

Oysters

Oyster mushrooms, especially *Pleurotus ostreatus*, are among the easiest mushrooms to grow. These aggressive mushrooms have a fast spawn run. In our indoor Shroom Room, under ideal conditions, we had a run in 16 days. Now, that's a fast turnaround! A number of strains are out there, and other species of oysters can be cultivated as well. We've grown Italian oyster (*Pleurotus pulmonarius*), pink oyster (*Pleurotus djamor*), king oyster (*Pleurotus eryngii*), and golden oyster (*Pleurotus cornucopiae* or *Pleurotus citrinopileatus*—one is from eastern Asia and the other from Europe). We will discuss all of these, but first let us focus on ostreatus.

Outdoors: Easy
Indoors: Easy

We have seen this mushroom grown on everything from coffee grounds to particle board. This aggressive mushroom will make its home on most any woody or pulpy substrate containing cellulose. It is easy to grow, has a delicate flavor, and is extremely versatile for cooking. If you are a hobbyist only, this is the one to start with, but if you are looking to add an income stream, wine cap serves you better. The oyster clusters are made up of many caps stacked one on top of the other, so they are very fragile, and transporting them to market requires extra care. Bumping them around or exposing them to high temperatures and humidity can quickly make your beautiful oysters look not so pretty. You will see these in grocery stores, where they either look like they're transforming into raisins or into some sort of slime. Even in the best of cases, they never ever look as beautiful and healthy as the ones you will grow, simply because they do not like the industrial grow methods most large operations utilize and they are not happy to travel. In fact, when we started selling them, a number of folks came up and said "Those aren't the same oysters they sell in the market are they?" When Dave explained why the local supermarket fare didn't look like these, people began to realize what a beauty this delicate mushroom truly is.

FIGURE 3.4. Oysters (*Pleurotus ostreatus*).

Pleurotus ostreatus is a versatile mushroom, available in a variety of colors from snow white or grey to a steely blue, and probably a host of hues in between. You can inoculate logs, stumps, natural fiber ropes, straw, newspaper, coffee grounds, cardboard, and fabric. Heck, we're not even sure how many different substrates this mushroom has grown on, which makes it great for experimentation. We have seen it occurring naturally, and if you find it and identify it, grab some, because we are not the only ones who like this mushroom. We have enthusiastically found beautiful specimens, only to find them chock full of little beetles.

Here are some things to remember when ordering your oyster spawn:

1 Get it from a reputable supplier.
2 If growing outdoors, get a strain that will perform well in your climate.
3 Ask if the strain does better on wood or straw. These are the two main strains out there, and you want to match up your substrate with their strain of spawn. However, we have to admit we have used spawn meant for straw in wood and vice versa without noticing huge differences. But, we did notice the mycelium run on the substrate was usually slower. This isn't usually a big deal, but if you are inoculating logs with straw spawn, your run won't be as aggressive as it could be, and if you happen to live in a hardwood forest, you may have another aggressive native that likes your log too, outcompeting your oyster!

Below are some growing methods for all oyster species. We cover non-ostreatus in sections that follow.

Oyster Growing Methods

We will cover the few methods (on wood, in straw bags, on rope) that we've used successfully, and what we found to be the best and why. We learned about oysters by growing them from straw kits first, and when we bought our property, we started growing them on wood. When we went into farming production, we had both wood and straw running, straw being the predominate producer for us because of its simplicity and our ability to estimate when the flush would be ready for market or restaurant, usually within an error margin of a few days. However, inoculating enough wood with oysters will ensure a long-term supply to complement your more short-term straw production.

Oysters on Wood

Wood is slower than straw; it takes more effort to create an inoculated log, stump, or totem, and the mycelial run takes longer. However, yields are usually bigger and you can have them for years, and the mushrooms tend to have a stronger, woodsier flavor than straw-grown oysters. Our first attempts were on stumps. We cut a number of trees on the property and left 1- or 2-foot stumps to inoculate with oysters. In our garden, two 16"–18" diameter maples had grown about four feet apart from each other. We inoculated one stump with an oyster strain called grey dove and the other with blue dolphin. A bench over the stumps and surrounding plantings provided enough shade for them, and, because there was easy access to water, we had a pretty good idea they would do fine there. One day during our second fall there, Dave went to sit down on the bench and saw grey mushrooms pushing up through the bench and peeking out from under it too. We harvested about 3–5 lbs. and had a great dinner. After a cold snap two weeks later, the beautiful blue ones made an appearance along with some grey stragglers. We were onto something here!

Before we dive too deeply into the methodology of growing oysters on wood, let's talk about the tree species we grew them on. We have grown oysters on poplar (yellow), maple (sugar and silver), beech, and ash. We prefer the harder woods like maple and beech, but ash is probably better used as a long-burning firewood. We once got a load of poplar, which is a more pithy wood than the denser maple and beech; we used the *drill and fill* method. While the mycelium run was incredibly fast and we got a fairly decent fruiting, the wood started crumbling after only one winter, whereas maple and beech logs will produce for a few years.

Oysters on Wood: Totems

Making totems is probably the easiest method for growing oysters on wood. When you are cutting up a tree for firewood, keep sections of any logs that have a 10"–20" diameter; you'll cut these into wheels for the totems. You will need a contractor's bag or some other means of covering the "totem."

Materials

- Oyster spawn appropriate for wood production—about 4 cups per totem
- Flat-cut hardwood wheels, 10"–20" diameter × 8"–14" length; 3–4 per totem
- Flat-cut hardwood wheels, 10"–20" diameter × 2"–4" length; 1 per totem
- Contractor's bag or other covering
- Gauze or cotton

Instructions

1 Cut and collect flat wood wheels; you'll need about four for each totem. Cut one thin (2"–4") slab for the top of each totem you wish to build.

2 Find a flat, stable surface to stack your inoculated wheels, ideally in an area that maintains a temperature of 55°–85°F and high humidity.

3 Place the bottom wheel in its spot.

4 Spread a handful of spawn, covering the wheel with about one cup of spawn.

5 Place your second wheel on top and repeat the process, spreading another handful of spawn on top, add another wheel, etc.

6 After spreading your last cup of spawn, place your thin wheel on top.

7 Cover the entire totem by placing in a contractor bag. Pull the contractor bag almost closed and hold in place with rope. Leave a fist-sized air hole at the top so that the totem can breathe. Stuff some cotton or gauze in the hole to allow gas exchange, but keep contaminants out.

The first year, we placed our totems under the porch, and we had a decent flush. Once we had the indoor grow room, we let them run for 3 months in the right temperature range with a humidity of 80 percent or more. After your totem has run, take it out and place it in a shady spot on your property. If you don't run your totems in a controlled area, build them where you will want them to remain. Important: If you live in an area with termites, you should place your outdoor totems on concrete block, not wooden pallets, as the lower tier is susceptible to termite invasion. You always want your experimental mushroom spaces and first tries to be located in a spot that a hose with a sprinkler can reach. Mushrooms love humid, wet places! Remember to always watch your days. We tried never to let three days go by without watering if the humidity dropped below 50 percent. Lack of sufficient moisture is the most common failure when growing mushrooms outdoors.

A Word on Climate

The sweet thing about Berglorbeer, our farm, was that it was on top of a mountain on the Eastern Continental Divide. The mountain caught weather from the west, from the Great Lakes, and from nor'easter storms of the East Coast. It is a humid, cool area. But we had a few summers of no summer, where it rarely got above 80°F the entire season. The variable weather taught us a lot. We got to see who likes it hot, cool, and humid, or who could handle somewhat drier conditions. A cool wet summer is great for maitake, shiitake, but not so good for pink oysters. A hot, dryer summer is good for one strain of shiitake, but not another. Talk to your spawn supplier about your conditions. Usually, they are looking for input on what works where, so document your results and stay in communication with them. Generally speaking, though, humidity is a wonderful thing, and a water source near your mushrooms is ideal. Water at night, with a sprinkler of some sort, to reduce your water usage. Just misting works great and doesn't waste water, plus, mister nozzles are fairly cheap.

When stacking your totem, utilize other stumps, a leveled pallet (or concrete block, if in termite-prone area), anything you can to stabilize your stack. This is *not* an inanimate object, but a living breathing, growing structure. Therefore, totems, we have found, want to topple over! The spaces in between your wheels can swell after a rain and the mycelium will expand. Therefore, if you do not stabilize it, the totem will topple over. Another tip: Tell the kids to leave them alone! Having three small children, we found out that some kids find it fun, funny, or downright hilarious to knock over a totem.

Keep in mind as you are cutting, that you will be stacking your wheels, so cut as straight as possible. When you place your totem out to fruit, level off the surface as best you can. We always tried to keep the bottom totem off the ground due to the fact we lived in a hardwood forest. Many other species of mushroom would love nothing more than to move into your totem. They will compete with the mushrooms you are growing and will reduce your flush and overall output.

We found totems easy to construct and—with some creative pallet placement by trees—fairly stable. The drawback we found was that beetles and other mycelium-loving critters would reach the mycelium and have themselves a feast, which of course is not beneficial to production. Every place is different though, and we have many friends who have had great success with this method. It is worth trying.

Oysters on Wood: Drilling and Filling Logs

You will find this method described in the shiitake section of this chapter, and the instructions for oysters are the same as shiitake. With oysters, you can probably widen your spacing out on a log, therefore stretching out your spawn, gaining more inoculated logs. But you should stick with the prescribed shiitake spacing at first until you get a feel for how aggressive a strain is in your area. We found that the more exotic oysters (pink, golden, Italian) were not as aggressive in our area as we had hoped.

DAVE'S TANGENTS
Gathering Wood ———————————

On a crisp day in October, with lead grey skies and a cold north wind, I head out to retrieve wood for shiitake cultivation. Beech, oak, or sugar maple will do, but I need a lot of 4-foot logs. Ah, the life of a mushroom farmer. Good thing one of my best friends is a lumberjack, even though he teases me about taking the twigs.

To grow a number of choice edible mushrooms, you must learn how to run a chainsaw. I cannot stress this enough! Years ago, when I was a weekend warrior, I picked up a brand spanking new chainsaw and set out to clear some trees for our garden and for firewood. A double oak was in the way of my newly conceived garden. I looked at the trunk leaning to the left, cut a quick notch, got on the back side, and started to saw. Soon the cracking and popping started. The tree started its descent and hit the ground with a resounding thump! Kristin ran

TOP LEFT: Oyster mushrooms primordia (pins) starting to form fruiting bodies (stipe and pileus).

TOP RIGHT:
Beautiful pink oyster mushrooms.

LEFT: Wild Chicken of the Woods glowing amidst lush moss.

annulus

TOP:
Wild gathered Chicken of the Woods.

BOTTOM LEFT:
Recently fruited oyster mushrooms.

BOTTOM RIGHT:
Wine cap showing annulus.

TOP LEFT:
Shiitake grow block, showing various stages of growth and parts.

TOP RIGHT:
Immature oyster mushroom cluster.

BOTTOM:
Basket of mushrooms ready for market.

TOP LEFT:
Reishi mushrooms.
Antler form.

TOP RIGHT:
Maitake fruiting, primordia
(left) and fruiting body
formation (right).

BOTTOM:
Shiitakes fruiting from
repurposed logs, many
years after inoculation.

TOP.
Horn of plenty, or
black chanterelles,
or black trumpets.

BOTTOM LEFT:
Many oyster clusters
nearing harvest.

BOTTOM RIGHT:
Strong mycelial run
(white throughout) in oyster
production bag.

TOP:
Pioppino growing indoors.

BOTTOM LEFT:
Reishi mushroom. Conk form.

BOTTOM RIGHT:
Large maitake clusters recently collected from the forest.

TOP LEFT:
Wine caps under
tomato plants, showing
rich burgundy caps.

TOP RIGHT:
Wine caps harvested
at ideal time.

BOTTOM:
Table presentation
with shiitake block.

TOP:
Let's see, what can we make from these fresh garden ingredients?

BOTTOM LEFT:
Dinner prep with maitake, leeks and wild rice.

BOTTOM RIGHT:
Cooking shiitake samples at a market.

out of our new house and yelled, "What the hell are you doing?" Proud of my Paul Bunyan ways, I triumphantly said, "Making a garden and getting firewood!" My job was only halfway done. I prepared to drop the remaining half in the same area as the first. I cut the same type of notch, started sawing, and the tree shuddered. The popping and cracking started. All of a sudden, the tree performed an ice skater's twist and started its earthward decent right toward my truck! I stood there helplessly with eyes tightly shut. As it fell, I heard the resounding thump, but no crunched metal. When I opened my eyes, I found that the top of the tree had brushed my tailgate, and the power line to our house was dancing. Once again my intrepid wife ran out and yelled, "David, what just happened?" My response was, "No worries dear, she landed right where I wanted her to!"

An 18" diameter red oak weighs close to two tons. Cutting trees is not child's play. So, I found help: my lumberjack friend, who not only allowed me to gather his twigs after a job, but also taught me how to properly and safely cut. For mushrooming supplies, I go to his yards and take the tops of the trees he's felled. This isn't a bad deal, as it only costs me a case of beer.

Having a lumberjack friend has provided me with a source of logs, but most importantly, he taught me how to run a saw. The importance of learning the Dutchman's cut and how to read a tree's mass cannot be overstated. Safety for you and your property are your first concerns. If you are a commercial grower, knowing how to cut allows you to carefully select your logs, cut them to the right length, and achieve a more uniform load. You will also be more careful with the logs than others would be about not bruising bark, or including split, diseased, or damaged logs in your load. Of course, you will also want to familiarize yourself with the tree species you need. Obtain a good tree identification guide if you need to brush up on your local deciduous species.

If you own property, figure out a cutting plan, as you can denude an acre pretty quickly. Your local service forester can help you with the plan. If you can locate a saw mill that will allow you to take leftovers, that helps. However, most big operations don't want some mushroomer running around their job site. Small individual contractors are easy to

FIGURE 3.5. Gill the Logger.

FIGURE 3.6. Logs in pickup.

get to know and work with. Talk with larger landowners in your area and see if they will let you cut; or, you could obtain a public lands cutting permit. Remember, you are not looking for lumber-value trees, so you will be leaving the more profitable trees. Thinning trees of lesser value helps the owner get more money when they harvest. Be very careful not to mar or tear up the trees they want to harvest later, as you will quickly lose your source of wood.

I know of a mushroom farmer who has his logs delivered, cut to length for 2 dollars per 4-foot log. In some areas this may work, but remember, buying logs will cut into your already slim profit margin. If you have the knowledge, time, and willingness to cut your own logs, you can save a good amount of money. If you can get a work party together for the drill and fill or have the ability to hire help, knocking out a bunch of inoculations in one day can also be extremely helpful. Being a one-man show, the chief bottle washer, public relations person, and everything in between, I never had this luxury. But, really, I enjoy seeing the process through from start to finish. Everyone is different, though. A grower's location, connections, financial situation, and chainsaw skills will be determining factors in how mushroom cultivation wood is gathered.

Oysters on Straw

Straw is a great medium for growing a number of mushrooms, and when you first get into production growing, straw is the way to go. You get a fast run, so you can basically plan your harvest almost to within a few days. Following is a great method for making "quick" mushrooms and great mulch, and it can done indoors or outdoors.

We learned a lot about pasteurization when we decided to sell and market our mushrooms. At first, we tried just soaking bales of straw and then inoculating. Dave read somewhere that all you had to do was fill up a tank (not heated), drop a bale in, put some concrete blocks on it, and let it soak for a couple of days. This method was not the greatest mushroom producer, but it did produce a strained back from trying to get a fully soaked bale of straw out of a tank! It has been known to work well in warmer climates, where the spawn runs faster. It also produced a myriad of other "mushrooms" and molds and ended up being a housing project for crickets, earwigs, slugs, and other critters we did not care to have around our mushrooms and garden.

Here is what we learned about pasteurization after a couple of years of trial and error: You need to pasteurize the straw in hot water. Pasteurizing straw for oyster cultivation requires you to *soak your chopped straw in 160°F water for an hour*. We use a turkey fryer burner as our heat source. Helpful hint: Build a stable, level platform to set the heavy barrel on. Use three to four tiers of concrete blocks, so that the barrel can sit on top and the flame is about 4" below the barrel. Our three-tier setup allowed us to insert the burner below and peek in to make sure the flame was still going. Why build up with concrete blocks? A gallon of water weighs 8.34 lbs. Multiply by 55 gallons, and you've got 458.7 lbs. of water to deal with. And that turkey fryer burner was not made to hold 450 plus pounds!

Equipment and Materials Needed
- Oyster spawn appropriate for straw production.
- Polyethylene sleeves for production or spawn bags with filter patches (patches with a filtered opening to allow air circulation) for grow kits.

- A way to seal the bags—twine for large sizes or vacuum sealer for small ones.
- Straw bales—*not hay*.
- Pasteurization equipment: a 55-gallon metal drum from the food industry (not industrial) and a heat source, such as a propane-heated turkey fryer base.
- Waterproof thermometer to hang in barrel (a candy thermometer works great).
- Heat-proof mesh bags to hold the straw submerged in the barrel.
- Clean wheelbarrow, preferably with a small hole in the bottom, to drain excess water from the pasteurized straw.
- Chipper/shredder. It is best if you can chop up your straw. This gives you more substrate surface area, and allows greater compaction within the bags.
- A tool to make holes in the bags. A 2 × 4 with arrowheads in it works great (see figure 3.8).

Instructions (with more detailed discussion following):

1 Fill the barrel with water and heat to 160°F.
2 Chop up your straw with a chipper/shredder.
3 Collect the straw into large, mesh laundry bags, preferably ones with sturdy drawstring straps that you can attach to the chipper/shredder to collect the straw.
4 Pasteurize your straw in 160°F water for at least one hour.
5 Carefully remove the straw from the barrel and empty it into a clean wheelbarrow with a small drainage hole in the bottom.
6 Prepare the grow bags by placing straw and then spawn in alternating layers, each time shaking the bag to more evenly distribute the spawn and compact it so you can get more in the bag.
7 Seal each bag, larger ones with bailing twine, smaller ones (such as the grow kits to sell) with a vacuum sealer.
8 Poke holes in each of the large production bags (but not the smaller ones with filter patches).
9 Hang the larger production bags in the shade near access to a watering system.

FIGURE 3.7a. Oyster Inoculation. Shroom Classroom participants pasteurizing straw.

FIGURE 3.7b. Dave unloading pasteurized straw.

FIGURE 3.7c. Shroom Classroom participants filling oyster bags with straw and spawn layers.

FIGURE 3.7d. Shroom Classroom participant sealing oyster bag.

Chopping Your Straw

You do this so that you can compact your bag and get the most bang for your buck from your straw and your spawn. But how to chop? Sit there with a pair of scissors? At first we ran over our straw with a lawnmower (and bagged it). After a while, we found a lot of contamination, so we bought a separate mower bag, one we didn't also cut grass with, and we would always blow or tap it out and then hang it in the shed until the next use. But when we really cranked it up in production for restaurants and three farmers markets, we purchased a chipper/shredder. The chipper/shredder is a dangerous, but very useful piece of equipment.

First, cut the baling twine of your bale, and save it. If you do not cut off the bailing twine you will break, plug up or destroy your equipment. Bundle the twine and use it later to tie up and hang your finished oyster bags, to save a buck or two. We found it most useful to abandon the wire cage method we read about. Instead, we went to a store and purchased the largest mesh laundry bags we could find, ones with solid straps—the heartiest straps they make is what you want. Wet straw has some heft to it, so you want a strap that won't break. We found some that had drawstring straps and these worked great. We could place the laundry bag over the outlet shoot of the chipper/shredder to catch the chopped straw, simply pull the drawstring shut and place the bag in the barrel. After we put four of these in the barrel, we placed a cinder block on top to keep the bags down for the hour. You will have to play around with this. You will want something between the block and the bags, otherwise the block (or other weight) will roll off to the bottom of the barrel. We used a metal disk (one made for riding on the snow) or a plastic tub. The laundry bags with straps also worked well because we could take a wooden rod and slip it into the water, slide it through the strap, and pull the bag out, letting it drain into the barrel before placing it in a plastic wheelbarrow with a cracked bottom. You may ask whaaat? When you live in a colder climate and are getting ready for shows or an order, putting your hands in hot, then cold then hot water will be miserable. You will figure out ways to keeps your hands as dry as possible. A cracked wheelbarrow allowed further drainage and cooling of your straw. You do *not* want a lot of water in your grow bags. You want your straw moist, not soaking wet with water in the bottom of the bag. If you have soggy straw and/or standing water in your bags, you will get much more mold contamination than if you put moist straw in without standing water. The plastic wheelbarrow can be wiped down with a mild cleaning agent to keep your area as clean as possible.

Preparing the Grow Bags

Your next step is to place your straw in polyethylene bag/sleeves, which you can purchase from a mushroom supplier. We have purchased long 4 mil sleeves and then custom cut and sealed the bottoms with a vacuum sealer. This helps save you money. You should, however, work with the industry standard first to understand how big, heavy, and cumbersome these can be, especially when you are cranking them out and have anywhere from a dozen to three dozen going at once. We received instructions on how to measure the amounts of spawn per amount of straw to use. After a while, we figured out how to eyeball it by using a handful of spawn per straw amount. We suggest that a novice measure and inoculate according to suppliers' measurements. In time, you will learn how much spawn works and that certain strains can be inoculated with a lesser amount, while others may actually need a heavier dose.

Place a few handfuls of straw in the bag and then some spawn, and give it the "shake 'n bake," shaking the whole bag. This is a "lite" inoculation rate. Now, compress this straw/spawn down. You will take two handfuls of straw and place them in the bag, add a handful (about ¾ to 1 cup) of spawn, add two more handfuls of straw and compress. Keep repeating this process until your bag is almost full. You can try sitting on the bag to compress it down as much as possible. Then twist the top, tying it off with your bailing twine. Next, you will want to hang your bag. We hung some of our bags outside in the shade where a hose could reach, and some we hung inside in the grow room. You can create a number of bags with different strains depending on your markets and whatever strains produce best for you.

Your bag is now almost ready for production, but it has to breathe! Originally, we would take an arrow with a broadhead on it and every 9"–12" down the length of the bag, poke a hole, then rotate the bag approximately 6" and, diagonal from the previous holes, we would make another row, so that the holes in the bag created a diamond shape. When we cranked up to production of many bags, we needed something that would do the job more quickly. We found a tool used in southeastern

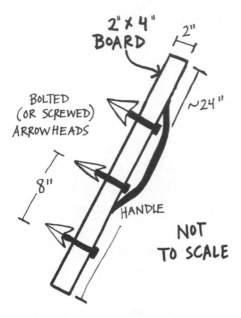

2" X 4"
BOARD

2"

BOLTED
(OR SCREWED)
ARROWHEADS

~24"

8"

HANDLE

NOT
TO SCALE

FIGURE 3.8. Oyster bag aeration tool.

Asian cultivation operations—a board with nails all through it and a handle. So we grabbed a 2 × 4, put a bunch of nails through it, added a handle, and poked the hell out of the bags with it. From that, we got very small clusters of mushrooms— not very marketable! So instead we took a 2-foot-long piece of 2 × 4 and drilled three holes 8" apart, put arrowheads through, bolted them tight and added a handle on the back side. With this, you could line up your board and poke holes in the bag quickly and efficiently! This tool is helpful, but it can be very dangerous, so be careful. We always hung it up on an eye hook in the corner of the grow room so that it was safe and secure and we always knew where it was. If you are not a hunter and don't have a readily available supply of arrowheads, simply go to a sporting goods store and purchase some (try the bargain bin first). You want solid arrowheads without folding blades.

Hanging and Watering the Bags

You should hang your finished large production bags in a shady area within reach of water. We used 2-gallon-per-minute misting nozzles above the bags when our humidity was below 60 percent. We would tie the nozzle above the bag using our extra bailing twine and slightly crack the faucet, just enough to create a mist around the bags. We would usually allow this to run for about an hour and then move it to another area with hanging bags. This is also very helpful when your bag is pinning and the humidity is under the 70 percent, helping you create bigger caps and clusters. You could also rig up a multi-mister system so that you do not have to move your hosing around. Use caution when watering. You do not want your mushrooms to become waterlogged! You want a humid environment, not a waterlogged one, which leads to quicker rot, softer mushrooms, and attacks by fungus flies.

You will have to experiment with adding water that nature has not provided to your mushrooms. You can try watering the bags for 15 minutes in the morning, 15 minutes in the heat of the day, and 15 minutes at sunset. This method works nicely for getting your young pins needed moisture yet allowing them to grow out firmly. When we are busy, and doing shiitake logs, maitake stumps, or getting ready for a delivery, we simply put the hose on, moved it around the best we could, and played the moisture level by feel. When you are in production outside, you will become very much aware of your temps and humidity. You can grow all varieties of oyster mushroom in this manner, which gives you a plethora of colors and flavors for your customers or your family. Oyster on straw can be grown outdoors or indoors. There are now compostable bags you can buy. They are a little more expensive, but we think they are worth it because they don't end up in the landfill.

Do not allow water used in pasteurization to just sit. When you're done pasteurizing the straw, you will need to drain the barrel so that the water and straw pieces do not sit and start to collect contamination (and reek). Don't just push over the barrel to drain it, though, as you will dent the barrel. Simply let your barrel cool after you are done, use a hose to gravity drain most of the water, dump what remains, and then hose the barrel out and let it dry for your next round.

We preferred using the round barrels so we could continue to use the lid, which keeps the heat in, helping with efficiency and therefore saving some money.

Oysters on Rope
We have had more fun working with rope inoculation than with any other method. We've had mixed results, but also had a ton of fun, and we learned a lot. The spawn run on rope seems to be a few days slower than straw, but faster than wood. You will need a natural fiber rope, no synthetics. We've used rope made of hemp, sisal, and jute. Hemp and jute seemed to work the best, but we never could find jute in an

appropriate diameter, so we just doubled or even tripled it up. Half-inch or ⅜-inch diameters work decently; you will infuse the rope with mycelium and place it in a cut you've made around a stump or log.

Instructions

1 Pasteurize your rope using the same method as for straw.
2 Place rope and spawn in spawn bags (we inoculated in 10' lengths, using ¾ cup of spawn for every 10').
3 Seal the bags.
4 Once the ropes are covered in mycelium, they are ready to use and at an ideal stage to sell.

If you are going to grow the mushrooms yourself:

1 Girdle a recently cut stump or larger diameter log (12" or more) slightly deeper than your rope's width.
2 Cut rope to length, if necessary.
3 Gently tap your rope into the cut.
4 "Seal" sites, as best you can, to protect them from "creepy crawlies."

Ideally, you will have a stump with a fairly thick diameter, 12" or more. Starting at the ground/stump interface, cut a groove (girdling an already cut tree), and then every 6"–12", repeat the process until you run out of stump. The closer together your inoculation grooves are, the faster your spawn will fruit. But your stump will break down quicker also, which can be a good or bad thing. It is up to you and what you want to accomplish. What we have found, similar to the totem method, is that a rope-spawned stump is a great feast for beetles, flies, larva, etc. The trick here is to seal your site the best way you possibly can. A few of the ways we tried were wax (which uses a lot of wax), newspaper stapled over the rope and covered with burlap, wax paper, and plastic (which did not let the spawn breathe enough for a solid run).

The best way to get a great return on your stump is to combine the methods: totem and drill and fill. With these combined methods, you can get a mushroom-covered stump. We prefer growing a single type of mushroom on one stump. But we have seen a mix-and-match approach where a couple types of mushrooms are growing from the same stump. As you become proficient in these methods you can get creative and start designing a landscape that reflects you. Some of our students tell us, "You have too much fun doing this." When it becomes fun, it's no longer work. But don't let anyone else know!

FIGURE 3.9a. Stump Inoculation Steps. Cut stump.

DAVE'S TANGENTS

Stump Experiment

One day, while looking at a stump of a tree we removed to get more sunlight on the garden and some firewood for the stove, I examined the stump and decided I was going to have fun and play around, hoping to create a project that would grow mushrooms and cover other bases as well. I removed three

FIGURE 3.9b. Spawn added.

FIGURE 3.9d. Rope spawn stump in winter.

FIGURE 3.9c. Bird bath stump.

wheels off the top, the first two were each approximately 4"–6" thick, and the very top one was 8" thick. I carved out the middle of the top wheel with chain strokes and made a rough bowl shape, and then I drilled four small holes through the bottom of my "bowl."

Next, I cross-hatched the other wheels with the saw and drilled more small holes randomly. Then, I girdled the bottom of the stump at ground level and every 6" until I reached the top, creating three girdles. I placed spawn ropes in the cuts, sealed them with wax, and then placed spawn in between the wheels, a la the totem method. Last, I placed the "bowl" on top. This became the "bird bath stump." When it rained, the bowl would fill with water, providing a place for woodland birds to bath, and then the water would slowly weep, or percolate, to the totem below, keeping all of the mycelium happy! After a rain, the bowl would hold water for 3–5 days, depending on temperature and humidity. You could see the totems, "weeping" a few days later! The thought behind this: a stump will still draw water up from its roots for a while, and the "trickle down" from the bird bath would keep the mycelium happy with moisture. The totem section fruited first, but beetles drilled through the wax to get some of the mycelium, so it didn't fruit for long. The rope section took longer, but produced better overall. The bowl was still intact four years later. Too bad the boy from *The Giving Tree* didn't have any mushroom spawn!

Other Oysters

You can experiment with the various species and strains of species of oyster mushrooms, using the methods already explained. Following are some other species you may want to try. Remember to pay attention to the species and strain, matching them with both your climate's growing conditions and your growing method. Communication with your spawn supplier is key. Also, experimentation and tracking your results can help you (and your supplier, if they welcome feedback on their strains).

Pink Oyster (*Pleurotus djamor*)
Outdoors/Indoors: Moderately Easy
(but cold sensitive)

Pink oyster is a fun and frustrating
mushroom to grow. It is an eye catcher,
with its brilliant pink colors and large
caps. Some say it has a flavor with a
hint of seafood, though we have never
experienced that flavor. Many people
love this delicate mushroom. Pink oyster is said to be very aggressive, and it
can be if its parameters are met. This
mushroom comes from a tropical cli-

FIGURE 3.10. Pink oyster (*Pleurotus djamor*).

mate, so keep that in mind. The pink oyster likes it hotter than some of
the other species we've grown. The temps should stay in the 70°+ range.
We found that nights below 55°F would slow down the spawn run and
primordia formation. While living in Pennsylvania, our outdoor pink
oyster production depended on our weather. During hot summers, we
had many beautiful clusters of pink oysters, but when we had cool and
wet summers, which happened frequently, we did not have very good
production. The straw became infected with green mold. It got so bad
one summer, we purchased cottonseed hulls to try as a new substrate.
The cottonseed hulls had a rate that was less than 50 percent infected,
whereas our straw was running 50–60 percent infected. During those
cool summers, we would usually get one flush and then the infection
of molds. We're not sure if it was the color or the smell, but one summer we had an infestation of beetles and flies on our pink oysters, while
other oyster species were growing nearby and were unaffected. During
hot summers, these issues never reared their ugly heads. We had great
production and happy customers. You should monitor your own climate. If your nights don't dip below 55°F, you should have beautiful
bouquets of pink oysters. If the nights get cold, you'll either have to
abandon the effort or move the bags inside. Due to the popularity of
this mushroom, we constantly tried to grow it for our customers. We

could always produce it indoors, where we controlled the temperature and humidity.

You want to harvest pink oysters before they start to drop their spores, because they start to develop a distinct, rancid smell when they sporulate. So, keep your eye on the fruits. You should look at the tops of the bottom caps. If they've started to sporulate, you will see white dust on the tops of the caps lower down in the cluster. Once harvested, pink oyster does not hold up as well as other mushrooms and must be handled with extra care. But they will move quickly at market, which is good because as they age after harvest, they start to develop a deep pink color and emit a foul odor, getting nasty sooner rather than later—not good for getting repeat customers! If your climate allows or you establish an indoor grow area, they are definitely worth trying to grow, as customers love them and they match up nicely with shrimp and a light white wine for a delectable pasta dish.

FIGURE 3.11. Italian oyster (*Pleurotus pulonarius*).

Italian Oyster (*Pleurotus pulmonarius*)

Outdoors/Indoors: Moderately Easy
Italian oyster is a hardier mushroom that has a crunch to it—if that is possible in a mushroom. There is a nuttiness to its flavor and it's probably one of the stronger-flavored oyster mushrooms. The Italian travels fairly well compared to the other species we've grown. We once grew pounds of this for a chef of a wonderful local restaurant. (It's so good, we got married there!) She incorporated the Italian oyster into a "Tour of Italy" dinner night. She absolutely loved this mushroom. Soon after the event, a number of her attendees were call-

ing and asking us to grow these mushrooms. The Italian oyster can tolerate cooler temperatures than the pink oyster. It fruits in the 45°–70°F+ range, so the cooler summers did not affect our production as much as it did the pink oysters. This mushroom is aggressive, close to, but not quite in the ostreatus range, at least in our climate. We found that with its thick caps, white stems, and dense clusters, this is a very fine, marketable mushroom that holds up well during travel, has a great flavor, and produces consistently.

Golden Oyster (*Pleurotus cornucopiae* and *P. citrinopileatus*)

Outdoors/Indoors: Moderately Easy (but cold sensitive)

The golden oysters are a beautiful yellow color, almost luminescent. They will catch eyes at market but, very much like pink oyster, they prefer a warmer climate—temperatures in the 70°–85°F range, which worked for us indoors. But, outside, they faired according the summer Mother Nature gave us each year. Of all the oyster species, we found this one to be the least aggressive in our temperate climate. Many times, we would have a great mycelial run, only to have it abort right after primordia formation. This never occurred indoors, only outside. We found that our cooler nights or cooler days affected production even after the spawn run was well underway. They are fun to grow, look great, and have a mild flavor. These are one of the more delicate mushrooms, but if you have a warmer climate, you may have great production.

King Oyster (*Pleurotus eryngii*)

Indoors: Moderately Easy

The king oyster is very different from the other oyster species we've grown. We found this mushroom to be our favorite for cream of mushroom soup, with its hearty stalks, and rich flavor. We grew this indoors on grow blocks and outside in a cottonseed hull/straw mix. For whatever reason, it never was a big draw at our markets. You need to know what people will and will not purchase at your market. Sometimes you just have to grow it, take it there, and see if it will sell. This mushroom has small caps and thick, fat stems that are entirely edible. We always felt it was their appearance that pushed potential customers away, at least in our market. But, it's a fairly hardy mushroom with a great taste and is not overly difficult to grow. It will tolerate cooler temps—in the 55°–75°F range—so it wasn't as susceptible to cooler summer nights as pinks and goldens were.

Other Mushrooms

Nameko (*Pholiota nameko*)
Outdoors: Moderately Difficult
Indoors: Moderately Easy
This is a delectable mushroom, but had no market whatsoever in our area. We couldn't give this mushroom away! Nameko translates to slimy mushroom, and yes, it has a gelatinous slime on it, but this disappears soon after hitting the frying pan. We would drill and fill nameko on cherry and beech. This is one of the few mushrooms we could use our plentiful cherry for. When we grew nameko on cherry wood, it did fairly well, but we saw no big difference in production between the two woods. Nameko loves moisture and doesn't mind cooler climates. Unlike shiitake methods, we would lay inoculated nameko logs on the ground in a hardwood chip "tub," an area dug out to approximately the depth of the logs and filled with the wood chips. This area was downhill from our shiitake stacks, so when we watered the shiitake, water would run downhill to the nameko bed. We really enjoyed growing this mushroom and love its flavor. But given that the slime made it unmarketable, we only grew it for our family use. It will fruit in the fall and you must be able to correctly identify this mushroom, as it looks like many native mushrooms, at least in the Eastern deciduous forest.

Pioppino or Black Poplar (*Agrocybe aegerita*)
Outdoors: Moderately Easy (in warmer climates); Moderately Difficult (in colder climates)
Indoors: Moderately Easy
After having grown this from a sawdust block grow kit, Dave simply fell in love with this mushroom! Its classic shape and incredible flavor won

him over immediately and therefore he had to grow it. Our lumberjack friend stopped by one day for beers and was talking at length about all of the poplar on a cut he was doing. Dave was at the cut site the next day, harvesting the unwanted scrap poplar to grow pioppino. We created beds by laying out four poplar logs as a box and then taking the chainsaw, scoring smaller logs and packing the cuts with sawdust spawn from our grow blocks. We also bought spawn, scarred poplar logs, inoculated them with the pioppino, and then buried them. We

FIGURE 3.12. Pioppino, or black poplar (*Agrocybe aegerita*).

always had a great spawn run, but never achieved the kind of fruiting we needed, as our beds would get overrun by some sort of green stinkhorn. We could grow this to production level indoors, but never could achieve what we wanted outdoors. Pioppino has a warmer spawn run, at about 80°F, and it will fruit at a cooler 50°–60°F. We believe anyone living in the eastern Pennsylvania valleys and southward, or in similar, more temperate climates, could produce this mushroom outdoors by drilling/filling and burying logs. Our outdoor temps never suited this mushroom, and what a shame, because it is Dave's favorite.

Lion's Mane or Hericium (*Hericium erinaceus*)
Outdoors: Difficult
Indoors: Moderately Difficult

Lion's mane is Kristin's favorite mushroom. Erinaceus is the more ball-like form of hericium, and we have found it in the wild. (Once, while bear hunting in northern Pennsylvania in late November, Dave looked high up in a tree and saw a basketball-sized hericium! Sadly, he and his buddy were deep in the woods, too far

FIGURE 3.13. Lion's mane, or hericium (*Hericium erinaceus*).

from a truck ladder or any other instrument to capture this massive specimen, which must have been 25 feet up.) We grew this mushroom outdoors and indoors. Outside, we would drill and fill oak and beech logs and then bury them upright to approximately one third of their lengths. We would get small baseball-sized fruitings from this method, whereas indoors, on sawdust grow blocks, we could get more clusters— many of which were volleyball sized. On the outdoor logs, we got one, maybe two fruitings, and these would occur October to late November. Much of what we read on lion's mane indicates that it fruits in the upper 60°F range, but in the wild, we would see it fruiting on sunny warm fall days, and there could even be snow traces on the shady sides of trees. Indoors, we always had better fruiting in the upper 50° to low 60°F range. This may just be attributed to the strain we used.

This mushroom is almost beyond description as far as taste. When sautéed lightly and a small pat of butter added right at the end, it has the texture and taste similar to crab legs. Recent studies have suggested that lion's mane can help with dementia and traumatic brain injuries. It soon became a favorite of a number of our customers; some liked its taste and others liked it for its medicinal properties. This is a recent phenomenon. When we first started selling it, only the bravest customers would buy it, as its appearance scared many off. But season after season, more people asked for it. Something, somewhere was written about it, or it was featured on a TV show, because it went from obscure to one of our most desired. As people become more educated about mushrooms, you will see more of this occurring, even with odd genera like *Hericium*.

Reishi (*Ganoderma lucidum*)
Outdoors: Moderately Difficult
Indoors: Moderately Easy
This conk-like mushroom is aggressive but has a slow spawn run. It grows on a number of tree species. We used to grow it on cherry logs in sand-filled pots. Much like lion's mane has come into vogue, reishi became one of our most sought-after mushrooms, fresh or dried, even in kits consisting of reishi logs placed in sand-filled grow pots. We could

sell out of reishi at almost every event we did. You can grow this mushroom on just about any hardwood, such as oaks, maple, and black cherry. We would cut the logs to about 18" in length, using those with a diameter that would fit into a 2.5 gallon pot filled with sand. You want the top of the logs to stick up out of the pot about 2"–4". We would place these under shelves in our greenhouse. This helped keep humidity and temperatures in a stable range. When reishi starts to form its primordia, you can change this mushroom's growth form by placing a plastic bag over the pot, elevating the CO_2 and creating the antler form, rather than the conk form. If you have reishi in the greenhouse with lots of plants, this may happen without the plastic bag. In Asian markets, these are sought after. If you have normal CO_2 levels, you will grow more of a conk form. We found that, in our markets, it simply did not matter. In fact, more people preferred the conk form in its entirety or sliced and bagged. You and your customers can make medicinal tea from this mushroom. It is not very tasty, but its health benefits are legendary in Asia. There are too many benefits to list here (see Chapter 7), but Dave can speak from personal experience that it brought down his borderline high blood pressure. Given his family history of high blood pressure, his doctor prescribed synthetic pills, which he hates. So, he tried reishi. After four months, his blood pressure was the best it had been in years. Each day, he drank an 8-ounce glass of reishi and raspberry tea, made according to the recipe found in Chapter 6.

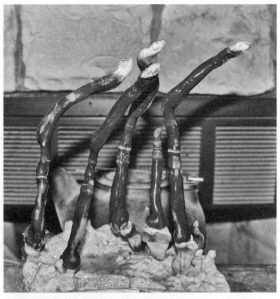

FIGURE 3.14a. Reishi (*Ganoderma lucidum*). Antler form.

FIGURE 3.14b. Conk form.

FIGURE 3.15. Maitake (*Grifola frondosa*).

The M & M's: Maitake (*Grifola frondosa*) and Morels (*Morchella* spp.)
We have tried relentlessly to grow both of these mushrooms—with limited success. Let us rephrase that…we had some success with maitake, but as for morels, we probably would have had better luck striking gold in Pennsylvania than cultivating a morel. We very successfully wild gathered maitake because, once conditions are right, this great mushroom will flourish, coming back year after year, sometimes in very large clusters. While we had some success wild gathering morels, they are more fickle than any cat ever put on this Earth, but that is part of their charm and mystique!

Maitake

Outdoors/Indoors: Moderately Difficult
Dave has picked maitake since his childhood. His Papap always referred to it as sheep's head, and it is also known as "hen of the woods." It can be found at the base of deciduous hardwoods, primarily oak, elm, and beech in the fall. Our best effort to cultivate it involved inoculating cut oak stumps. But our production never met expectations, nor could it rival what we found by just roaming the woods. We're not sure if production would've increased with more time spent on it. One stump

we inoculated produced two small clusters of maitake a mere six years later. Maybe, in time, it will continue and produce larger clusters. We have tried to grow maitake indoors also, which would result in some massive clubs but not the large fronds desired. We believe there was a CO_2 issue that caused this growth pattern, but are not 100 percent sure why this occurred. We have met with and talked to other growers who have had great success with this mushroom—but they had issues with mushrooms we could grow easily. It just goes to show that every setting is different. As you gain growing prowess, it never hurts to talk with others. Learning should only stop when they close the lid on the coffin!

These pictures on the right show maitake stump inoculation.

Black, Yellow, and Grey Morels (**Morchella angusticeps, M. esculenta** *and* **M. deliciosa**)
Outdoors/Indoors: Difficult (very difficult)
Every spring for a decade, we would purchase morel spawn and set forth to create our very own super-secret morel patch. After a decade of trying, we had absolutely no success whatsoever! We would put spawn around apple trees, would burn an area and put down spawn (*angusticeps*), turn in shovels full of compost and put down spawn. If there was a surefire method, we have tried it. Every time we planted a fruit tree, we added morel spawn, just in case. After a decade, we have come to the conclusion that our soil probably would

FIGURE 3.16a. Maitake stump inoculation steps. Measure and mark holes in cambium (outer) layer.

FIGURE 3.16b. Drill holes.

FIGURE 3.16c. Inoculate, seal holes, and label stump.

never produce morels. We had some limited success wild foraging on our mountain, but when we headed east into the limestone-laden ridge and valley province of Pennsylvania, that was a different story all together, as was the very southwestern corner of the state and the Shenandoah Valley and James River in Virginia. All produced banner morel days. We feel that soil and geology have more to do with your capabilities for cultivating morels than anything else. The morel is a mysterious mushroom, probably the most sought after, and has more names and "surefire ways to find them" then all the rest of the mushrooms combined. The morel, its legend and lore, is a book itself. You don't need to stop trying to cultivate it or locate it, as this is part of the fun. Also, *do not* get frustrated if you are selling mushrooms and many folks only ask for morels. It's going to happen. Of course, they will ask for white button and portabella also, which is when we simply direct them to the nearest grocery store, as we never had any urge to grow those compost-based mushrooms. For the some of the latest research on indoor and outdoor morel cultivation, see *Organic Mushroom Farming and Mycoremediation* (Cotter, 2014).

Shiitake (*Lentinula edodes*)
Outdoors/Indoors: Moderately Easy
This is the mushroom that got the cultivation bug kicking in Dave's soul. At one time, he was content wild gathering hen of the woods, morels, chanterelles, and papinkis (*Armillaria mellea*), the honey mushroom. Of course, his fresh mushroom consumption was limited to the spring, late summer and fall. Dave is forever grateful to his friend Jody (who became his brother-in-law) for introducing him to shiitake growing. Dave would receive grow kits and logs after visits to Jody's home in Florida. Jody started cultivating shiitake as a hobby. When Jody started growing more and more shiitake, he went into business. From hanging grow kits in the shower to watching in horror as Florida hardwoods turned to pulp after one or two winters, Dave decided to dive into growing his own shiitake. When we purchased our first home in Pennsylvania, his growing took on a life of its own. The first thing Dave did was selectively

FIGURE 3.17. Shiitake (*Lentinula edodes*).

drop trees for next year's firewood and purchase all of the necessary tools for our first serious drill and fill. For years, the "mushroom yard" was just beyond our garden, within reach of the hose, in a nice, shady spot. Dave was on his way.

Cultivating shiitake is a labor of love—there is no way around it. To drill and fill logs and then to nurture and care for those logs while awaiting your first flush, perseverance is required. The fastest we have ever seen a shiitake log go from inoculation to cultivation was about eight months. You will have to learn the strains and how long they take, what their parameters are, as these factors, and more, have great influence on when your logs will produce, how much they will produce, and how long they have to rest. These many variables are learned in time, both in general terms and also according to your unique environment.

Dave quickly fell in love with cold strains of shiitake, with their thick, beautiful caps. But cold strains took *at least* a year from inoculation to fruiting. We sometimes had to wait 18 months, but were usually rewarded with a spring and fall flushing from that point on. Like cultivation of other mushrooms, shiitakes can be fickle. But once you figure it out, you will probably find that they are relatively simple to keep in cultivation. Until that time, it's just part of the journey of a mushroom cultivator.

FIGURE 3.18. Shiitakes popping from our logs.

Shiitakes on Wood

The first thing you will need is a source of wood. Oaks are the most preferred of all the trees, as they will produce for years before breaking down. "Spent" oak logs can then be used for landscaping or creating raised beds (see Chapter 5 for details on this and other sustainable methods). Beech and maple were our next most-utilized species. They both produce nicely and aren't as heavy to move around as oak, but their bark will break down quicker than the rough, thick bark of oak. The real key to your logs is what's inside. Lumberjacks look for big heartwood; shiitake growers are the opposite. We want a small heartwood and a thick sapwood/cambium layer, (in cross section, that's the part of the log beneath the bark and surrounding the heartwood). Fast-growing, straight saplings that need to be thinned are perfect. We found that tree tops from a logging site worked nicely, especially since this is "garbage" or "scrap" wood in the timber industry. You want to keep your diameters where you can handle them fairly easily, with 6"–8" diameters working best for shiitakes and many other mushroom species grown on logs. The straighter the log the better, as you will be stacking these logs, so curves, limbs, and other growth patterns that deviate from straight make stacking harder—increasing your *mayhem*. A side note: We would also inoculate curved and funky logs. When we were selling inoculated logs, these seemed to be preferred for their "artsy" appeal. Or, did the customer think that because the log was funky, the mushroom would be too?

The best time to cut trees is in the winter, when they are dormant. This allows for more sugars in the sapwood and the bark is tighter than at any other time of the year. We have cut and inoculated trees throughout most of the year, but late fall, when the leaves have dropped, until spring, before the buds break, is optimal. Live trees cut in the spring through fall have sap flowing up or down. Also, many actively growing trees produce antifungal compounds, as they fight existing mushrooms in their environment. For these reasons, the dormant season is your best bet for collecting wood for shiitake. If you come across wood that is available during other parts of the year, don't necessarily pass it by. If it is still "green" when you get it, before inoculating those logs, simply let them sit for two weeks or more. It is also important that, no matter when you harvest your logs, you make sure they do not dry out! This will have an adverse effect on your results. Logs waiting to be inoculated should be covered to protect them from wind and sun.

Strains of Shiitake

Shiitake's increasing popularity has led to it being grown all over. Because of the demand, a number of suppliers have developed strains that perform better in different climates. For the beginner, we suggest starting small by obtaining small amounts of plug spawn of a couple of different strains, and then figuring out which work best for you and your area. Talk with suppliers and see what they suggest for your climate and available materials. Some strains are said to do better on red oak or white oak, for instance. Also ask about the length of the spawn run, which is the time it will take from inoculation until primordia formation. Some run in as little as six months, while others can take up to two years. You will want to know these details. The primary types of strains are cold weather, warm weather, and wide range. Their names are self-explanatory but within these categories you will find a variety of named strains. We used all three strains because we wanted a prolonged fruiting season starting in early spring and continuing through late fall. You can force a fruiting of shiitake by soaking the log in cold water over night and then whacking the end of the log with a hammer a few times. We have done this during the heat of the summer, when our logs are least productive and our farmers markets are at their peak, but we prefer to allow nature to run its course. We feel it's healthier for a log's life and harvests are better when conditions are naturally right for fruiting.

All the strains have tendencies that you will have to learn, some producing better in your area than others. We often had conversations with our brother-in-law in Florida who was also growing shiitake. We found that some strains that were trouble in Florida produced like gangbusters in Pennsylvania and vice versa. If you are just starting out and can only afford one strain, we suggest that the wide range strains are best for the beginner. They tend to do fairly well most anywhere. All of the wide range strains we used performed pretty well for us. Cold weather strains were our favorites, probably because they fruited first in the spring and the mushroom caps were beautiful (they had darker colors and were thicker). But they take longer to run. Warm weather strains seemed to cause the most consternation in our area. Some did very well, and others had problems, spotty runs, and low production. But this was probably due to our weather patterns and temperatures, not the quality of the strains.

We never really noticed any differences, as others have, between red and white oak. But, we primarily grew on red oak, so we may not have had enough experience with white. As for beech and other woods, wide range strains seemed to do the best, but cold strains worked well for us also. Beech breaks down quicker than oak so, when using beech, we stuck to wide range because cold strains take longer to run. Talk with your supplier, telling them where you live and the tree species you plan to use. They want you to be successful so that you will buy more spawn and other supplies from them in the future. Most of them supply across North America, so they will have some insight to share. If they don't, you probably want to work with another supplier. A number of garden catalogs now offer "shiitake plugs," etc. We have talked to many people at fairs and markets that have had limited success with these, probably because these businesses don't focus solely on mushroom spawn, which is a complex and highly scientific business. Your best bet is to work with and order from a reputable supplier who is willing to help you out along the way. We ordered spawn from a few different suppliers and found that the one closest to us supplied us with the most successful spawn. Perhaps this is a coincidence, but we feel that it may have to do with similar growing conditions and therefore more suitable strains. See our list of suppliers in Resources.

Inoculating Shiitake Logs: "Drill and Fill"

There are many ways to get shiitake spawn into a log. We will name and explain a few, but we'll focus on the one method we found to be best overall, in terms of production, time, ease, and labor. You can take a chainsaw, cut out a wedge, pack with spawn, and replace the wedge, anchoring it with nails or screws. This may be the oldest method, but we found the spawn dried out, bugs would bore in and eat the mycelium, and native mushrooms would colonize parts of the log, creating competition. We have also tried rope spawn with shiitake, but it never bore fruit for us. We found that the rope method works better with an aggressive species like oyster.

The most common method and the one we had the most success with is *drill and fill*. This is where you drill holes and then fill them with sawdust spawn, plugs (small dowels covered in mycelium), or pre-designed plugs with a Styrofoam cap, called a "thimble." We have used all three of the available spawn types and found sawdust spawn to be the most economical and successful, followed by the dowels, and finally the Styrofoam-capped ones (they had a tendency to lose their caps over time, drying out the spawn). The dowels are effective, and we have used them successfully with other species. But when inoculating 200+ logs a year, we found the sawdust to produce the best and be the most economical.

A higher speed drill works best, and your drill bit needs to match the method you are using. The best bit we found was one that has a stop collar on it that matched up in diameter with the inoculating tool. The stop collar ensures that you drill to the right depth each time, so that you don't have to guess how deep to go.

Though there are a lot of different drilling patterns out there, we found that a diamond-shaped pattern was the easiest to lay out and drill. It also had great spawn run success. If you are looking to knock out a bunch of logs, the diamond pattern is the way to go. We also did a spiral pattern, and when the log fruited, it was very unique looking. We would do a couple of patterns that became mushroom art, which were used as eye catchers at shows. When you work hard, you should also have some fun. So, even though 99 percent of our logs were straight, diamond-patterned production logs, we also had a small collection of weirdly shaped logs with short branches and different inoculation patterns, but these were just for show.

Instructions

1 Place a log on your work station.
2 Take the tape measure and marking crayon to lay out your diamond drilling pattern. (After stealing our kids' sidewalk chalk, Dave discovered that tree marking pens, merely oversized crayons, work best and are relatively inexpensive. Plus the kids don't get upset that their favorite color of chalk is gone. These can be found where chainsaws and similar equipment are found.)
3 Stretch the tape the length of the log, securing the claw at the far end.

Approximate Hole Spacing for Various Species: Hericium—4"; Reishi and Nameko—6"; Shiitake—6" to 8"; Oyster—10"

4 Down the length of the log, mark the first row of hole locations with your crayon, about 6" to 8" apart. Once you have figured how aggressive your strain is, or you are drilling and filling other species, you can tighten this measurement or widen it out.
5 Once you have your first row laid out, pick two holes on either end of the log, put your tape measure dead center and measure 2" on either side of the marked holes and make a mark. Remember to mark both ends of the log and on either side of the existing row. Rotate your log and lay your tape on the marks you made. You will again be making marks every 6" to 8", but you will want them to

Materials

- Shiitake spawn: sawdust, plug, or thimble
- Oak or other appropriate hardwood logs—approximately 4' in length × 3"–10" in diameter
- Tree marking crayon (or sidewalk chalk)
- Tape measure with metal claw on the end
- Drill and drill bit
- Inoculating tool (for sawdust spawn)
- Wax (not needed if using thimble spawn)
- Wax bowl (not needed if using thimble spawn)
- Wax daubers (not needed if using thimble spawn)
- Equipment to heat the wax, such as a propane stove or turkey fryer (not needed if using thimble spawn)
- A surface suitable for drilling logs (two saw horses with boards to place logs on)
- Aluminum log labels and ballpoint pen
- Heavy-duty staple gun
- Pallets
- Pine mulch (optimal, if your yard is in a hardwood forest)

be made 2" off the original row, and dead center between the existing holes. When you repeat this on the side, you will see the diamond coming into form. You will continue to rotate and mark holes in this alternating pattern around the log. If you have a branch coming out or some kind of anomaly, you can add an additional hole or two. We would also add a hole or two around the ends, but we would never put a hole less than an inch from the edge of the log.

6 Drill holes in the log at each mark, ensuring that you have the right depth each time (the same depth as your inoculating tool).

7 After you have created holes around and up and down your logs, you are ready to insert the spawn into the log. If you are

 Lesson Learned: One cold winter's day, Dave was inoculating logs with a brass tool, smacking it with his palm. He soon realized spawn was escaping everywhere *except* into the hole because he had cracked the brass. He can be a bull in a china shop! So, he quickly got on the phone and ordered a more hardy stainless steel version.

using an inoculating tool, you simply pack as much spawn into the tool's cavity as you can, place over the hole and push down with your thumb, or smack it with

FIGURE 3.19a. Shiitake log inoculation materials. Log, ruler, and marking pen.

FIGURE 3.19b. Spawn and inoculation tool.

FIGURE 3.19c. Wax dauber.

your palm. This will push a plug of spawn into the hole. You will repeat this process until all of your holes are filled with spawn.

8 After you have the holes filled with spawn, double check to make sure you have them all filled. When you first start, do a "finger test," pushing down on the spawn with your finger. Next, you will seal the hole with wax. There are a number of kinds of wax and ways to apply it. When we were hobbyists, we used tub wax, a soft wax you pinch and place over the hole and press down with your thumb. This wax works ok, but it is messy and when it gets warm in the summer, we found the wax would stick to our clothes and hands when we were moving logs around. Cheese wax became our standard, which needs to be melted and applied over each hole with some sort of instrument that is dipped in the hot wax. Be careful, as this wax can catch on fire! For application, we found that wax daubers, little cotton balls on a piece of twisted wire, work the best and are fairly cheap. You can purchase these at mushroom spawn suppliers. You simply dip the cotton end in your melted wax, and then just dab wax over your holes.

9 After you have covered your inoculation sites with wax, double check you have covered them all and that they are completely covered with wax. Sometimes you will see a little spawn peeking out, so you will need to apply more until the spawn is completely covered. An opening to the spawn is an opening for creepy crawlies who eat the spawn and mycelium you worked hard to insert. You don't want to feed anyone who hasn't helped! Openings also invite other mushrooms to colonize the log.

10 Label each log with an aluminum tag on the end. Record the date and the strain of mushroom, etching the information with a ball-point pen. Attach the label with a sturdy staple gun.

11 After you have inoculated all your logs, stack them in your pre-chosen shady area with water access. A conifer stand works best, as no competing hardwood mushrooms are present. We like to put pallets down and stack the logs on top of them. First, we add pine chips under the pallet to help cut down on native hardwood

fungus attacking the logs. When you stack your logs, lay down a row, leaving space between the logs for air circulation. Place each successive row perpendicular to the last one so that when finished, your stack is cross-hatched. Try to place like-diameter logs together if you have a mix of sizes. Place larger diameter logs on the ends and smaller ones in the middle to help with air circulation. Usually each row had between 4 and 7 logs, depending on diameter. You can stack them as high as you want, but we like to keep our stacks between 4 and 6 rows high. This way we can look into the stack and see if any primordia (mushroom pinning) is occurring.

12 You will want to stand your logs up when it is fruiting time because big stacks can be cumbersome to harvest from. You can lean logs against almost any kind of structure, like a shed or trees, or you can construct custom log stands from lumber.

13 Remember to keep your logs moist when they are fruiting. Mist them at night if there is no rain in the forecast. An hour of misting is usually sufficient.

When we inoculated in the dead of winter, we would bring those logs indoors, keeping them in our basement for a couple of weeks or until the outside temp was above freezing. This helps speed up your spawn

Helpful Tip: Especially if you are producing shiitakes for sale, you will want to keep a notebook, so that when logs begin to fruit, you can look at the tag, mark how long the spawn run took, and how much was produced. This helps you narrow down which strains do best in your area and identify any problems with your methods. It can even help you determine your profit margin and pricing, especially if you record how much time you spend on inoculation. You can also note temperatures, shape, color, cap thickness, and other characteristics of the mushrooms, once they fruit. Again this will help you narrow your work to strains you like and produce what you are looking for. To keep your information organized, you might want to create and print inoculation log spreadsheets and keep them in a three-ring binder near your work area, such as in a nearby shed with your other equipment.

FIGURE 3.20a. Shiitake log inoculation steps. Student marking drill holes.

FIGURE 3.20b. Student drilling holes.

FIGURE 3.20c. Student filling holes with spawn (note diamond pattern).

FIGURE 3.20d. Dipping the dauber in hot wax.

FIGURE 3.20e. Student sealing inoculated log.

FIGURE 3.20f. Labeled shiitake logs stacked in cross-hatch arrangement on pallets.

run. Just remember that houses can get very dry in winter. Make sure you keep your humidity up if you are going to do this.

Indoor Cultivation: The Mayhem Within

Indoor cultivation—what can we say about it? We could go with "Cleanliness Is Next To Godliness," but just as relevant would be "Insanity is often the logic of an accurate mind overtasked," (Oliver Wendell Holmes, Sr.). Or, we could tell the tale of Sisyphus. Or we could remind you of the deceitfulness in your own mind—that you are in control. Indoor cultivation can and will become a necessary evil/experiment/exercise if you are going to seriously produce mushrooms. Unless you live in a climate that allows you to consistently produce a wide variety of mushrooms outdoors, you will undoubtedly dabble in the art of mushroom growing in its highest form: indoor cultivation. Simply put, an indoor grow space allows you to create and maintain optimum cultivation conditions throughout the year.

Everyone with whom we have discussed indoor growing has gone through the roller coaster of outcomes. It is part of the mystique of

FIGURE 3.21. Shiitakes in the Shroom Room, our indoor growing space.

indoor cultivation. You will probably open the door to your grow room one day and experience the lowest of lows. We do not want to discourage people from trying, just to be prepared. We hope to help you avoid some of the humps, bumps, and pitfalls of indoor cultivation—or at least lessen the pain when it comes. The Number One rule of thumb with indoor growing is this: *Cleanliness Is Next To Godliness*. Repeat, repeat, repeat. Unless you have the money, time, and energy to create a "clean room" that would pass laboratory or hospital specs, be prepared for incidents of contamination. As Hunter S. Thompson said, "If you're going to be crazy, you have to get paid for it or else you're going to be locked up." One of the only reasons to grow indoors is to make money— you will be able grow mushrooms with an indoor grow room like no one else can, at whatever time of the year. But growing indoors also allows you to understand mushroom parameters better, and hopefully you can project what you learn to a grander scale. We are in the mushroom growing business, so indoor cultivation is a necessity—evil, crazy, the root of the mayhem, but also very fun, educational, perplexing, and rewarding. Without further ado and build up, let's dive inside.

Creating the Grow Room

First, assess what you have and how you can economically create an indoor grow room. We have seen elaborate "new" rooms created in basements, complete with professional carpentry, portable greenhouses brought inside, large plastic storage boxes the size of steamer trunks, and custom outbuildings created, from the start, for mushroom growing. Whatever works for you and your capabilities, we will get you started on your road to indoor cultivation.

However you construct your grow room, you want to control contaminants as best as possible, be able to create and keep humidity, control temperature, and be able to move around in there when you have to work. You will need air circulation and the ability to bring in filtered fresh air, if possible. You want your walls, floors, and ceiling to be of a material you can clean thoroughly. If your room is not squeaky clean, it will become the growing medium, especially if you grow an aggressive mushroom, like oyster. We saw pictures from a customer of ours who used plywood sheets and 2 × 4s to wall off a section of his basement. He grew oyster mushrooms in it for about a year and half before he realized that his grow room itself was growing! Spores from the oyster had colonized the walls and studs, and mushrooms were popping out everywhere.

We decided to convert an 8' × 6' coal bin into our grow room. We looked at the coal bin, and, after much deliberation and a cost assessment, we busted out a power washer and blasted all the black coal dust off the walls, floor, etc. We then siliconed the ceiling (which was the floor of the room above). We purchased marine quality white paint and began painting the cinder blocks that made up the walls and the ceiling. Lesson learned: cinder block can suck up *huge* quantities of paint—and marine paint is expensive. If we had a redo button, we would have used a cheaper white paint for the first two coats and then finished with marine paint. Plywood won't suck up as much paint, so you may want to assess what you can and can't build with from a financial aspect. Stiff plastic walls and ceiling work well because plastic is easily kept clean.

After the paint was on, we set out to create as sterile a room as possible. We put a fan in the coal chute window to the outside and cranked a heater on. Our humidity was low—in the high teens—so we basically baked the room for three days. The next step was to figure out how to get fresh air in without contaminants. We took an air filter (hepa is best), created a framework for it, and silicone-sealed it to the concrete chute. On the room side of the filter, we installed a small window fan, which we could use to push air out, pull it in, or both. With our project, we had to fill the auger hole (which was where coal was fed into the coal furnace on the other side of the wall) with foam sealant so that contaminants did not find their way in through the hole. Since the room was a coal bin, the door was small, starting at about 2" above ground level. It featured three Plexiglas pieces you could slide on a track to see in but basically keep air flow from outside the room to a minimum. We purchased two shelf racks for growing blocks on. (Do yourself a favor and buy the metal kind, not the cheaper plastic ones. The metal ones are much easier to wipe down and have better air circulation.) We anchored two 1" dowel rods to the walls, low enough that we could get our hands around them, and we used these to hang large oyster grow bags. If securing rods right next to a wall, make sure that they are built far enough away from the walls that the bags hang free (about 12" is enough clearance). Similarly, adjacent parallel rods should be at least 24" apart from one another. We could fit 6–8 bags in the room. We found, though, that 6 was the optimal number, as this allowed enough space for oyster clusters to grow out of the bags without touching one another.

You will have to purchase a few things to make your own room a working grow area. You need a hygrometer, which measures temperature and humidity. We have seen these run as high as $200+ and less than $40 from a garden supplier. When you are getting into business, you should cross-pollinate your needs. As your peruse all of those gar-

FIGURE 3.22. Progression of oyster mushroom growth in our Shroom Room (all three pictures depict the same oysters).

dening/greenhouse catalogs, think not only of your gardening needs but also about your mushroom growing needs. You can find economical ways to get your supplies by looking in a variety of places and online. You will also want a humidifier in your grow room, one with a fine, misty spray. You want your grow room humid, but not wet. Standing water should never be in your grow room. This is the host of enemy number one: green mold. We found an ancient humidifier in the back of our local hardware store. This thing was as big as a small dresser, held two gallons of water, had a filter and was awesome. We had used some of the smaller, cheap ones, but we were constantly adding water. This wasn't ideal because you want to avoid entering the room as much as possible to help reduce contamination. Also, opening the door on certain days can drop the humidity a surprising amount, sometimes by 10 percent in less than a minute! You will also want a small heater and another small fan to prevent stagnant air (we put ours opposite of our air exchange, which made it very effective). You should walk into a breezy, warm, mushroom-aroma-filled oasis when everything is clicking, rejuvenating you as you enter. Of course, when you look in and see green mold, it will elicit a completely different response.

Understanding the Indoor Grow Room

Before you purchase or create grow blocks, study your room and your mushrooms. This may actually help you decide what you should grow indoors. It will also help you figure out how to manipulate what you want to grow. Right outside of your grow room, you will want a clipboard with your cultivation logbook. In it, you will have a couple of columns: outside temp, outside humidity, inside temp, inside humidity. You should operate your room for a week or two before growing inside, monitoring how your room is reacting to your manipulations. This will help you get control of the grow room, and allow you to plan ahead if your room reacts to storms, fronts, or Mother Nature's other twists and turns.

After you manipulate and monitor the room for a while, you'll find out where it seems to settle and level out at optimum conditions. This will give you great insight. Shiitake like a temperature in the 65°–80°F range, but cold weather strains prefer slightly cooler temps, down to 55°F, and warm weather strains prefer temps closer to 90°F. Keeping your humidity between 75–85 percent works great. Anytime we got into the upper 90 percentiles, we had condensation, which is no good. With shiitake, when you notice that your grow blocks are ready for primordia formation or you want to force primordia formation, you will need to drop your temperature by a minimum of 10°F. We usually tried to drop the temp closer to 20°F because we found the closer we got to that 20°F drop, the more uniform all the blocks responded. With a 10°F drop, some blocks would start and others would lag behind, by up to four days.

Oysters preferred temperatures in the 75°–80°F window, with pink demanding closer to 90°F. Reishi liked the 70s, while hericium (lion's mane) liked it cooler than all the rest, with 70°F being its happiest level. We grew pioppino on the shoulders of fall and spring and could grow it with hericium. We could successfully grow cold weather strain shiitake with hericium in late fall through winter and into spring. We could also grow warm weather strain shiitake with bags of oysters. This is why it is so important to understand your room and the mushrooms you want to grow. Maitake is one of Dave's favorite mushrooms, but we never had real success growing it indoors. It would start strong but then "club up,"

FIGURE 3.23. Pioppino and lion's mane in Shroom Room.

not forming the fronds. We believe that this had to do with CO_2 levels, as our antlered reishi let us know that this was probably a good hypothesis. Reishi will grow in its antlered form with elevated CO_2; with more normal levels it will conk.

Once you understand your room and its reactions, you will need to decide which mushrooms you want to grow. Very much like outdoor cultivation, indoor growing has its quirks, and some species are easier to grow than others. Many suppliers offer grow blocks, so a lot of the work is already done for you. These blocks are augmented sawdust blocks, inoculated with spawn. You simply need to provide them with the right conditions, and they will produce mushrooms. All the reputable suppliers send their blocks with instructions, so you should start with their guidelines. We suggest starting with one type of mushroom, maybe two strains, to see which performs better. As you build confidence, you can start adding "like-minded" mushrooms, ones whose parameters overlap and can grow in the same temp/humidity windows. Happily, we found out early on that shiitake grew easily in our room, which was always profitable.

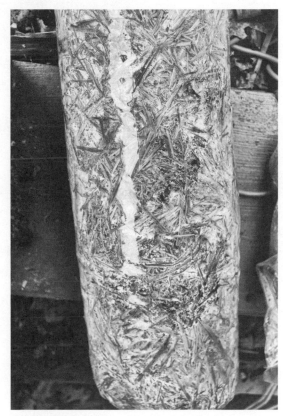

Scourges and Plagues

Find a mushroom cultivator who has grown indoors for any period of time and whisper "green mold" in his ear. Be prepared for blood curdling screams, a smack upside the head, or a shaking, quaking ball of humanity at your feet sobbing uncontrollably. Sure, we may be exaggerating the dangers in this introduction to indoor growing, but infections are a serious matter indoors. The importance of a clean grow room with good air circulation cannot be overstated. Molds, black and green, cause you to shut down and sterilize your room all over

FIGURE 3.24. Oyster production bag with green mold invasion.

again. Another common contaminant is the fungus fly, who we think always packs a small haversack of green mold on their backs. Even when everything is running smoothly, though, it's extremely important that you find some time to rest your room, giving it a complete wipe down. Be sure to replace filters and clean floors, fans, shelves, and walls regularly. When you enter the room, you should have clean clothes on. Try your best to keep a clean work area and be mindful that spores are everywhere. Still, there can be challenges.

Once, we had a big special order coming up and needed to have the indoor grow room produce like a machine. It had been running all summer without a hitch, putting out hericium and cold strain shiitake like a champ. We had even produced pioppino and Italian oyster for a restaurant's special event. We were gearing it up for this new special order, when they hit: fungus flies and then green mold. We should have shut down the room for a day or two and thoroughly cleaned everything. But we were rushing it and had left an oyster grow bag in the room when we added the new blocks for our big order. We didn't pay close enough attention or really inspect the room. We soon saw some fungus flies zipping around, but didn't see any green mold. A few days later, we saw some specks of green mold, and when we got into the grow room, we saw that the bag left in there had a sizable band of green mold on the backside. Luckily, we were able to salvage enough of the kits and grow them outside. We owe Mother Nature a thank you note for the mild autumn that allowed us to fill that special order.

For more outdoor and indoor growing methods, see the cultivation books listed in Resources.

Wild Collecting
and Purveying

Wild purveying is the collection of wild mushrooms for sale to customers. *Collecting*, to Dave, means "getting dinner." An avid outdoorsman, Dave values hunting and fishing, which play a large role in our family's diet, much as they did with our peasant ancestors. We were all hunter/gatherers at one time. To Dave, searching for mushrooms is more like hunting than gathering. When hunting mushrooms, you learn the symbiosis between trees and mushrooms or fire and mushrooms (morels), and much like hunting game, you start your search for mushrooms by reading the habitat, not just by walking through the woods looking for mushrooms. If you just walk in the woods, you will see mushrooms, but chances are they won't be the ones you are seeking.

"There are old mushroom hunters, and bold mushroom hunters, but there are no old, bold mushroom hunters." —AN OLD SAYING

Learning where they live is part of the battle, but much like a white-tailed deer hides from hunters, it seems mushrooms have an ability to silently sit there camouflaged and let you walk by. Or worse yet, you can be looking right at a mushroom and not see it. It happens, and it's not just morels that have this ability. Black chanterelles (black trumpets or horn of plenty) are notorious sneaks, blending perfectly with the decomposing debris on the forest floor. Once, Dave was staring right at a 2-foot-wide maitake (hen of the woods) and was about to pass on, when our youngest child Sage yelled out, "Look at the size of that mushroom!" The sad part was that Dave was looking for hen of the woods. The same thing will happen to you, and you'll simply have to laugh at yourself. Other mushrooms are just brazen and laugh at the world, like chicken of the woods, or they play cruel tricks on you, like the basketball-sized hericium (lion's mane) that Dave saw while bear hunting; it was two miles in and 25 feet up in the air!

If you decide that wild gathering is something you want to do, you need to make sure you know what you are doing. Better yet, find someone who knows and join a mushroom club. See Resources to find local mushroom clubs. Even if you know what you're doing, if you are selling wild-collected mushrooms, you have to know the rules of that game and be mindful of the fact that most people have never tried many species or may not wish to try them if wild collected. We always clearly state when we are selling a wild-collected product so that wary people can make a decision to pass, if they wish. Dave always tells everyone we sell mushrooms to, cultivated or wild collected, to always try a little bit at first. Everyone reacts to new foods differently. Even the most hardened mushroom connoisseur can lay down in pain after eating three pounds of fresh morels at once. Shaggy manes (inky caps) are a well-known mushroom for the table, but if you consume alcohol with them, you might be doubled over. We used to sell a handsome quantity of chicken of the woods at our markets. We would post pictures on our Facebook page of our collection, and people would be waiting anxiously for them. After hearing some stories of upset

FIGURE 4.1. Dave with hen of the woods (maitake).

stomachs from wild collectors of chicken, we figured out that they all had one thing in common: they plucked the chicken of the woods from black cherry trees. Wilted black cherry leaves have a cyanide component, and wounds on a cherry tree also have some cyanide, probably to help seal and cure. We know that chicken of the woods grows on wounds, so those mushrooms may have had some cyanide in them. Mushrooms are like sponges, so we never sell chicken of the woods from black cherry trees. We have never had a single complaint or a stomach-ached customer of our chicken of the woods—or any other mushroom we've sold.

Always know where your mushrooms come from. We once had a great customer, who was a very intelligent man. He told us how he had eaten some wild-collected king boletes and felt awful that night. We had never heard of anyone getting sickened by king bolete before, and we know he had eaten them many times. When Dave asked where he picked them, the lightbulb came on as the words came out of his mouth; "Golf course, I was golfing at…" We both had a pretty hearty laugh at his expense. He still searches for boletes, just not on golf courses. We can't tell you how many times we've had people bring pictures of hen of the woods or other mushrooms they picked in their neighbor's yard while walking, and asked, "Can I eat that?" After asking the owner, "Do you mind if I pick that?" your very next question should be "Do you have your lawn treated? Do you use glyphosate herbicide or other treatments?" If they do, *DO NOT pick and eat!* You wouldn't have a nice plate of morels and chase them with a shot of glyphosate, would you? We suppose you wouldn't, even if you're an advocate of the chemical's safety. So, *never* eat mushrooms from places that "treat" for this or that—bugs and weeds being the most common "scourges." In the next chapter, you will read about sustainable methods and how we never use sprays or treatments on our property. Sure, we get dandelions, but we've also had praying mantids, toads, spiders, and a garter snake who liked to scare the crap out of Dave when he went to pick beans. The more diverse your landscape, the fewer issues and problems you will have with various kinds of pest outbreaks. Monocultures get diseases and pests quicker, and they spread faster. You must be aware of your surroundings, even in the woods, along roadways, and at parks, etc. Some states spray park areas, so be conscious of these things.

Every state or province has its own laws and regulations concerning mushroom gathering and purveying. Federal lands have numerous regulations and laws. Even within each state or province, some areas or types of lands will have different rules and regulations. For instance, you may not be permitted to gather in a state park, but may be allowed to within state forests. It is up to you to know your local, state, and federal rules on wild collecting *and* purveying. Claiming ignorance will not get you out of trouble. We cannot stress enough the importance of 100 percent sure identification and 100 percent sure compliance. Your mushroom club and your state or provincial departments of agriculture should be able to provide you with the information you need.

Nature is the source of all true knowledge. She has her own logic, her own laws, she has no effect without cause nor invention without necessity. —LEONARDO DA VINCI

Wild collecting mushrooms is one thing. Selling them is a whole additional issue. In some jurisdictions, if you are gathering for yourself,

FIGURE 4.2. Wild reishi on very old beech tree stump along a stream.

FIGURE 4.3. Dave with wild-gathered reishi.

you may just need just a wild collection permit for public lands—or nothing at all. But if you are selling them, you may also need a commercial wild collection permit. In many locations, a Pandora's box of permits, procedures, and certifications is being opened widely. Wild collecting, and even purveying, are gaining in popularity, so laws and enforcement are soon to follow, especially with the recent focus on food safety, as demonstrated by the US Food Safety Modernization Act. For your own sanity and protection, learn what is and is not legal. This will save you and your business from headaches you simply do not need. This chapter is not an identification guide, nor a license granting you permission to collect wild mushrooms. However, we've developed some tenets for wild collection and wild purveying that will help you start your wild mushrooming journey safely and happily.

Tenets to Follow When
Collecting Wild Mushrooms

1 Be safe in the woods.
2 Always be 100 percent sure that you know what you've collected and that it's a safe mushroom to consume, taking into account its unique growing environment, i.e., type of tree (so, know your trees, too), and environmental health.
3 Start with easy-to-identify mushrooms, and work your way on from there. We started with chicken of the woods and others Dave was already familiar with, then moved on to boletes, which are more difficult to positively ID.
4 Learn everything you need to know about 100 percent positive identification using reputable resources, such as lifetime wild collectors in your area, who will likely be members or leaders of local mushroom clubs. Go out with these people to learn in a hands-on fashion; use the identification resources he or she recommends.
5 Don't just go online when searching for identification resources to use and assume that all resources are accurate and that you are capable of keying out all mushrooms with 100 percent confidence—unless you're a trained mycologist.

6   Collect only fresh, just right, specimens that have not yet released spores and do not have any rot or major bug infestations.

7   Leave some be, only collecting some of the mushrooms present, thereby promoting future regeneration through sporulation.

8   Be aware of property boundaries.

9   Don't collect from any areas where the public is prohibited from collecting wild mushrooms.

10   Don't collect from any areas that may have been chemically treated, such as golf courses, lawns, and roadsides.

11   Don't collect from any areas where there is a likelihood of heavy air pollution.

12   Know when a wild collection permit is needed on any state, provincial, or federal land, and know their collection rules, such as take limits.

13   Always have permission to collect wild mushrooms from private lands.

14   When consuming new mushrooms, always try just a little bit at first.

15   Always cook cultivated and wild-collected mushrooms thoroughly.

FIGURE 4.4. Dave with chicken of the woods mother lode.

Tenets to Follow When *Purveying* Wild-collected Mushrooms

1   Follow all of the tenets for wild collection above and ensure that your customers do as well, plus…

2   Always let potential customers know when your products are wild collected, giving them the freedom of choice when it comes to trusting you and the safety of your wild collection. We used laminated signage on our display board for each of our products, so this is where we let customers know.

3 Always know whether a "commercial" wild collection permit is needed to collect and sell mushrooms from public lands.

4 Know whether your state or province limits the sale of wild-collected mushrooms to certain species.

5 Know whether your state or province limits the sale of products to ones from the business property *and* whether any other public or private property ownership rules apply.

6 If it's ok to collect and sell mushrooms from private property, have the owner's permission in writing.

7 Be aware that liability, insurance, and legal matters may be involved.

8 Consider taking a wild-collection course, if possible, to ensure that you are a safe purveyor and to engender trust in you.

9 Know whether your state or province requires a wild-collection certification for purveyors of wild-collected mushrooms. This is a growing trend, so be on the lookout for this requirement in your area.

Every now and again we will hear through the mycelial mat (the grape-vine, for us mushroom folk) that someone got poisoned picking chanterelles—specifically *Cantharellus cibarius,* because they picked "the mother lode," only it was really jack-o-lantern, *Omphalaotus olearius.*

Dave has never understood how this could happen. Chanterelles usually grow more scattered, whereas jack-o-lanterns grow in big clusters on stumps. What's more, chanterelles have ridges and jacks have gills. Dave always saw the bright orange of a jack-o-lantern from very far away, and never saw a chanterelle *that* orange. If ever in doubt, simply take a question-able jack-o-lantern into a dark room. In a few minutes, it will glow, and its luminescence will allow you to see its

FIGURE 4.5. Display board signage for chicken of the woods.

gills. If you are unsure, wrap it in wax paper, keep indoors, wait a couple of hours, and repeat. Chanterelles never glow, and if you are ever unsure, and we mean not 100 percent sure of your identification, *do not eat*. This is just one example of the confusion that can spring up with wild mushrooms. You need to know what you are picking, and not just be sort of sure.

DAVE'S TANGENTS — Childhood Wild Picking Adventures

I have always had mushrooms in my life. They are woven into my very fabric. I was born in Johnstown, Pennsylvania. My maternal grandfather (Papap) Martin Vrtis was pure Slovak, having immigrated to Canada in 1937. During my childhood, Papap was retired and always willing to take his grandchildren on adventures. Considering that there were five children in the Sewak household, it's pretty certain that my mom (Violet) was relieved to lose a child or two for a few hours to the care of her father. Papap's excursions could consist of country drives across covered bridges (including a stop for ice cream!), fishing at a local sportsmen's pond, blackberry picking, or mushroom hunting.

It's the mushroom picking method he used that I want to share, from the vantage point of four plus decades of experience. Papap would drive to a spot he had marked as probably having mushrooms, possibly during hunting season, when he noticed it. He would pull over and get out, handing each kid a pocket knife and a basket. He carried a walking stick and a giant basket with lots of paper grocery bags. Nana (Judith) would also have a big basket, a walking stick, and a babushka (handkerchief tied over the hair). Martin would lead the troop down the path, instructing us to look for tree stumps and mushrooms.

He would have us scan for mushrooms on oak stumps when we were out to harvest pipinki (*Armillaria mellea*), most commonly referred to as the honey mushroom. Once a cluster of these was spotted, Papap would go over to inspect the mushroom. He liked the caps a pinkish brown color, and an annulus was paramount. He preferred them before the veil broke, but he wanted to see at least one with a broken veil and an annulus, in order to make a positive ID. Once he determined that the cluster of mushrooms was indeed pipinki, then he would get out a bag, cut the mushrooms, tap them, blow on them, pick off any forest-floor litter like leaves, larger bugs, and slugs, and place his treasure in one of his paper bags.

While he collected, the rest of us would fan out, but stay in sight of

FIGURE 4.6. The Sewak kids with their chicken of the woods find.

Papap. This way, greater ground was covered. It also explains why, to this day, I love to climb and explore mountains, rarely staying on the path. A lot of times someone would find hen of the woods, maitake (*Grifola frondosa*). This was always a special find. I still remember Papap walking over to the mushroom and digging at the base of the stump. His favorite part was the blob of mycelial mass just under the ground. Hen of the woods can be a substantial find.

Nana and Papap would make a giant pot of maitake soup, which I called "Nana's cleaning-out-the-fridge soup." The actual recipe doesn't exist and probably never did. She would submerse a maitake in cold water with the fronds facing the bottom, gently gyrate them, and then pour out the liquids with any leftover debris. A number of pots would then be filled up, the quantity depending on the size of the hen or if a number of them had been found. Nana would chop up the hen into small pieces, then go to the fridge and start chopping whatever she had into the pots. Some had meat (bacon type) or something else; celery, carrots, onions, and cauliflower would be in some pots, but not others. She would make three or more versions of soup and freeze it for the winter. They were all different and yet all the same…good!

Because of Dave's childhood experience wild collecting, he has taken all three of our children mushroom hunting, passing on his knowledge to them. By having multiple sets of eyes scanning the forest floor, the stumps, and the trees, you greatly increase your chances of finding good mushrooms. Having eyes scan at different levels is also a bonus. Once, Dave and the three children were out looking for chicken of the woods. During a rest period, our youngest Sage asked, "Dad what are these black mushrooms?" Soon the whole clan realized they were in the middle of a horn of plenty (black chanterelle/black trumpet [*Craterellus fallax*]) patch. Gently, all four of us picked, and within an hour and a half of scanning the hillside, every bag was filled. Seventeen pounds of the freshest mushrooms! These little black trumpets have a great ability to hide in decaying debris of beech clusters. From Dave's higher vantage point, they never stood out because the trumpets point upward, much like last year's beech nut husks. Thanks to a 5 year old and his lower stature, we found them.

To find this mushroom and see why it is called horn of plenty, look for areas where a beech thicket gives way to a mixed forest. In a pure beech stand the numbers won't be as great as they are in the transition zone. Usually, where another branch/stump of a different species crosses one of the beech sprouts (saplings) from the roots of the mother tree, horn of plenty can be found. Beech are very interesting, and there is a symbiosis between beech and the horn. When the gang goes looking for the horns, look for the biggest beech tree, and spread out, all walking slowly, in a zig-zag pattern toward it, looking at the ground for the trumpets.

This also brings up the need to be able to identify trees. So many connections exist between species of trees and the mushrooms

FIGURE 4.7. Sage and Cassidy Sewak collecting horn of plenty.

you will find on or under them. Every mushroom hunter should carry reputable mushroom ID books and good, local tree ID guides. Learn how to use them and make mental or written notes, so you can remember where you found your mushrooms. A listing of both mushroom and tree ID guides can be found in Resources.

In our Shroom Classroom how-to classes, we handed out a simple booklet, which we have included in our appendix. This booklet contains a glossary of terms and information

FIGURE 4.8. Horn of plenty, or black chanterelles, or black trumpets.

on cultivated and wild-collected mushrooms. More importantly, we would provide an optional end-of-day mushroom walk in the forest around our house. Almost all of our students, most of whom had long drives ahead of them, chose to participate in these walks. We would arm them with our own collection of field guides on trees and mushrooms. As we walked, we talked more about relationships than specifics. Once we got in an area with mushrooms (which Dave had scouted ahead of time) we would narrow down specifically what we were looking for. An example: for one of our fall classes Dave had located a maitake (hen of the woods), so we walked up a trail to that area, approximately a quarter of a mile. In our booklet, we had this description and a picture:

Grifola frondosa (hen of the woods, sheep head, maitake)
Wild—This excellent mushroom is found in the fall, look around oak stumps. "Hens" are very hard to see initially, but once found will usually produce every year at the same stump! Cool nights and autumn rains bring them out.

The students were all in a semi-circle around Dave as he asked the students with the tree books what they were noticing. We happened to be in a beech transition zone in an area where oaks were logged years ago. Previously, when scouting, Dave had spotted black trumpets, jack-o-lantern, maitake, and some LBMs (little brown mushrooms—there are so dang many of those out there—we simply look, see "LBM" and move on). Quickly, one of the students saw the jack-o-lanterns, which were past their prime. When one of the students was in an area where Dave knew the trumpets were, he asked the tree folks to identify the trees, which of course were mixed beech and oak. It took a while, but finally someone noticed the trumpets. The elation and joy of the whole class was overwhelming. A few took a couple of handfuls to take home and try. After some time discussing bolete identification and a bunch more LBMs were found, Dave asked if we had found them all. Everyone scanned around once more, and they were convinced we had found all mushrooms in the area. Dave asked them to read the description on maitake we provided as he walked up to the hen sitting within 20 feet of our group (you can see this specimen on our cutting board in Chapter 6). Heads were shaking in disbelief at this final discovery.

The short of wild gathering: Go with someone who *knows*, follow our tenets, and join a local club. When you do join a club, we advise you to make a donation. If you don't want to donate or don't have available funds for full membership, think of volunteering to help out with the website, the newsletter, or whatever brings your strengths to the table and helps the club. It will be greatly appreciated, and can help your and others' knowledge grow. You will find that mushroom people are very much like mushrooms; possibly, Kristin and Dave are like the more plentiful black trumpets, whereas others are morels, with degrees in mycology. What we all have in common is a passion for the world of mushrooms and spreading our spores (knowledge), perpetuating mushroom gathering, and cultivation for the next generation.

Companion Planting and Other Sustainable Methods

We hear the word everywhere these days: sustainability. But what does it mean? How does it differ from resilience, a concept gaining in popularity (for good reason). How does it relate to other similar principles, like permaculture? Why is it important? How does it apply to mushroom growing?

Sustainability is the endurance of systems and processes contained within, or the ability of systems to sustain their processes over time. So we should take a quick look at systems thinking before proceeding to sustainable mushroom cultivation.

Systems Thinking: It's Not Just About the Mushrooms

"What?" you say, "But it's all about the mushrooms!" Die-hard enthusiasts tend to view the world through blinders, no matter what their passion, whether it's fly fishing, mushrooming, or some other awesome part of this wonderful world. It's natural for a person to become a specialist, as many species in nature do. This way, the individual can excel at filling a unique niche and more easily outcompete others for that niche, while other niches are filled by other individuals or species. While this focused approach is very much a part of the modern human experience, as seen in our professions, the planet is in decline, and once-resilient systems are breaking down. It's not just about overpopulation, oil drilling, or deforestation. It's life-supporting systems that are breaking down. How is your mushroom cultivation venture going to strengthen the systems within your sphere of influence, such as your economic situation, the health of your gardens, or even your spiritual and physical well-being?

What's the Difference? Economics vs. Ecology

We believe that a fundamental flaw that leads to the breakdown of life-supporting systems—whether they are natural or man-made—lies in not connecting economics and ecology. The root of both words is the same, *eco*, which originates from the Greek word *oikos* (spelled in English here), which means *house*. The health and resilience of our economy depends upon the health and resilience of our ecology. Why? Because they are basically the same thing. Every form of business, directly or indirectly, depends upon natural resources in many ways. The health of our natural home impacts the health of our economy, just as the sustainability of our business practices affect our ecological health. They are interconnected, interdependent. Unfortunately, much of Western civilization views them as very different, even opposing, phenomena, creating a serious disconnect that is unhealthy for our environment and will eventually have grave consequences for economies as well. Having a foundational understanding of this interdependence can help you to better believe in, understand, incorporate, and communicate your own sustainable practices in life and in business.

What's the Difference? Sustainability vs. Resilience

The root word *sustain* implies remaining the same, so the root focus of the sustainability movement is to do no harm—meaning anything that would prevent conditions from remaining at a desired state of being. It's traditionally been about not causing further harm, that is, conserving systems and processes that are still in a healthy state, preserving the balance and equilibrium within. Like resilience, sustainability is partially about endurance. However, resilience, at least in the context of systems, whether natural or man-made, attempts to go a step further, challenging the notion that systems are in states of equilibrium. Rather, systems are in a constant process of evolution and change, for better or worse. Change is both a natural function of systems and a phenomenon purposefully brought about by manipulation. Like sustainability, resilience is a system's inherent ability for endurance, but unlike sustainability, it's about adapting, even when facing transformation, evolution, or significant change, not just during a state of equilibrium, balance, or non-disturbance. It's sort of like the difference between surviving and thriving. In today's volatile physical, economic, and social environments, it pays for us to give some attention to the evolution of systems, transformative forces beyond our control, and our own desired changes. Therefore, we should consider incorporating sustainable practices, but also keep the lens of resilience in focus so that we and our systems are better able to adapt to emerging changes from outside sources and shape the transformation we hope to see.

Biomimicry

Coined by Janine Benyus (1997), biomimicry is derived from the Greek *bios* or *life* and *mimesis* or *imitation*, and has three components:

1 Nature as model: Observing and imitating or taking inspiration from nature's functional models. An example is Velcro®, which mimics the burdock plant to achieve its adhesive qualities.

"Look deep into nature, and then you will understand everything better."—ALBERT EINSTEIN

2 Nature as measure: Evaluating our innovations as resilient or not, right or wrong, according to ecological standards, as nature has developed what works and discarded what doesn't over 3.8 billion years of evolution. For example, nature does not produce any non-biodegradable waste. Therefore, the movement toward reduced and eventually zero waste is using nature as measure.

3 Nature as mentor: The philosophy of valuing nature for its inherent wisdom, rather than just for its extractable and non-renewable resources; so it's going beyond learning *about* nature toward learning

from nature. It's about ushering in a new era, a new paradigm of resilience by learning from nature and applying its principles, rather than working against nature.

In Giles Hutchins's work on biomimicry for business (2013), Fritjof Capra's "Principles of Nature" are explained so that we may emulate them in business and in life, according to the biomimicry philosophy.

1 **Networks:** Just as mycelium is a network of cells and collaborates with the outside world, nature is full of examples of networks—systems of interconnected, interacting, sometimes interdependent organisms and resources.

2 **Cycles:** The flows of energy and matter, like water and nutrients, result in no waste, but continually transform into other forms that are usable by others.

3 **Solar energy:** On Earth, the sun is the basis of most energy. This nearly limitless resource is the foundation for most life forms and their ecological cycles by fueling photosynthesis in green plants, which is the capture and transformation of chemical energy.

4 **Partnership:** Related to networks, partnerships and collaborations are formed for mutual benefit so that a diversity of niche-fulfilling organisms can specialize but still derive needed benefits from their partners.

5 **Diversity:** In natural systems, the more biodiversity they incorporate, the more resilient they are.

6 **Dynamic balance:** The same as "dynamic equilibrium." Systems are not just in balance or a state of equilibrium; they are continually in motion, with various components changing, responding, and adjusting for the resilience of the overall system.

Permaculture Principles

A very useful sustainable design philosophy is *permaculture*, which, according to one of its two primary pioneers, Bill Mollison (1991) is "a philosophy of working with, rather than against nature; of protracted and thoughtful observation rather than protracted and thoughtless labor; and of looking at plants and animals in all their functions, rather than treating any area as a single product system." Mollison's partner in pioneering permaculture, David Holmgren (2002), developed the following 12 guiding principles, based upon the three foundational ethics of Earth care, people care, and fair share.

1 **Observe and interact:** Taking different perspectives, learning from nature. Two-way communication.

2 **Catch and store energy:** When resources are abundant, collect and store them for later use.

3 **Obtain a yield:** Produce as much harvest as possible, but also enjoy your yield for immediate gratification.

4 **Apply self-regulation and accept feedback:** Work with the ecosystem, not against it. Care for the entire system. Adjust your approach according to the feedback you receive.

5 **Use and value renewable resources and services:** Favor the use of renewable resources from nature over consumptive uses and expensive technology.

6 **Produce no waste:** Recycle and reuse to get the most out of the system and your investments into it.

7 **Design from patterns to details:** Step back and observe natural patterns, then focus on the details.

8 **Integrate rather than segregate:** Make an effort to create mutually beneficial relationships because the whole is greater and stronger than the sum of its parts. This is where permaculture's focus on creating polycultures comes from. A polyculture is a community of food organisms growing together. It includes the concept of companion planting and also relates to valuing diversity.

9 **Use small and slow solutions:** Smaller systems use less energy and are easier to maintain, and thoughtful, steady changes are usually the best solutions.

10 **Use and value diversity:** Biodiversity in natural systems ensures resilience against changes in the environment.

11 **Use edges and value the marginal:** Intersections are where the most interesting, diverse, and sometimes valuable phenomena occur.

12 **Creatively use and respond to change:** Change is inevitable, so learn to adapt to it and even anticipate it; use it to your advantage.

Putting It All Together: Resilient Design

Sustainability. Resilience. Permaculture. Biomimicry. What are the primary tenets each of these philosophies share and how can you incorporate them into a design for your landscape and gardens to make them resilient in all ways—ecologically, economically, and socially? The first step to sustainably growing mushrooms is to assess your site and the resources available. If you're committed to a healthy environment, leaving your property in better shape than you found it, or to a resilient system, you should stand back and look at your site as a whole system, one in which mushrooms will play a beneficial role. Coined by mushroom guru Paul Stamets (2000), *mycotopia* is "an environment wherein ecological equilibrium is enhanced through the judicious use of fungi for the betterment of all lifeforms." We suspect that he was thinking of dynamic equilibrium, balanced systems that evolve and can handle change because of their resilience.

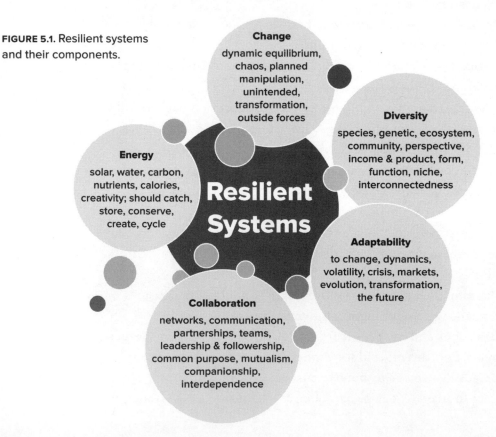

FIGURE 5.1. Resilient systems and their components.

Change
dynamic equilibrium, chaos, planned manipulation, unintended, transformation, outside forces

Diversity
species, genetic, ecosystem, community, perspective, income & product, form, function, niche, interconnectedness

Energy
solar, water, carbon, nutrients, calories, creativity; should catch, store, conserve, create, cycle

Resilient Systems

Adaptability
to change, dynamics, volatility, crisis, markets, evolution, transformation, the future

Collaboration
networks, communication, partnerships, teams, leadership & followership, common purpose, mutualism, companionship, interdependence

Follow these steps to create a resilient design:

1 Determine your property's assets. Where do you have access to pure, non-chlorinated water? Where do you already have (or want) gardens? Where are there (or could there be) trees or other plants for shade? Sure, you want to be thinking mushrooms, but if you have a blank canvas, think of *systems*, such as landscapes around your home that are edible or a natural forest garden rather than a mapped-out flower bed. For example, you could install a diversity of beautification species, vegetable plants, and wine cap mushrooms in a mulched bed.

FIGURE 5.2. Use rain barrels to water your mushrooms.

2 List the goals for your property based upon your overarching goals, your property's assets, and your desire to incorporate resilient elements. What components do you want to include and how is each one using a sustainable method? Look at the permaculture principles and the figure "Resilient Systems" to generate your goals, asking questions like:

- How do you want to transform your property?
- How are you cycling, rather than wasting, energy within your system?
- How are you capturing and using solar energy?
- How are your companion plantings going to create interdependence among the species paired together?
- Would you like to be able to water your shiitake logs with rain barrels—gravity-feeding captured rainwater to them?
- Do you want to create wildlife habitat and grow food for your family?
- How is your business diversified and creating interdependent partnerships?

Look through the examples that follow for ideas about sustainable mushroom growing, but also be creative. There are countless possibilities, so use your imagination.

3 Draw your property from an aerial view, depicting its current condition. Don't worry, it doesn't need to look fancy. Use a regular pencil for current conditions. Then add features you'd like to in-

corporate, also in regular pencil, which allows for alterations. Once you have everything where you'd like it to be, color in your wish list items, using colored pencils. Your landscape design drawing does not need to be exact or have perfect dimensions at this stage. However, it is helpful to have a good idea of scale, slope, sunlight, and other major considerations when drawing new landscape features and certainly when installing them. For example, you need to be able to estimate things like how many stacks of shiitake logs will fit in a shady spot with good water access.

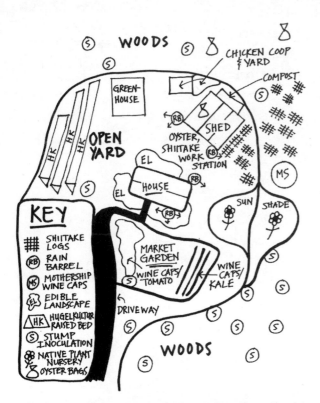

FIGURE 5.3. Landscape design from Berglorbeer Farma. Though it may help with the beauty of your design on paper, you need not be a landscape architect to thoughtfully design a resilient landscape that incorporates mushrooms, as Stamets says, "for the betterment of all lifeforms."

4 Prioritize your wish list and come up with a phased timeline for each upgrade. You won't be able to do everything at once, so this will help you make changes in manageable steps. If you want to make profit on shiitakes within two years, you must include shiitake log inoculation and placement within year one. Your plan is not set in stone, though. You have the freedom and power to continually adapt and make changes.

5 If you are growing mushrooms as part of a business, you will want to develop your business (see Chapter 8) and landscape plans at the same time, as there is overlap between the two.

Applications for Mushroom Growing

Yes, it's true that systems thinking and a resilient landscape with mushrooms playing a critical role are the ultimate goals. But we understand how difficult it can be to think all at once about the entire system—its many, many components, and the infinite ways in which resilience may be achieved. It's like asking you to write, read, and talk all at the same time. Multi-tasking is, in most cases, overrated, and we would even argue that it's counterproductive. So, we would like to offer some examples and ideas about techniques that could be used within your resilient systems. Many more examples exist than are discussed here (including from your own creativity). However, here are some examples of specific components or techniques we've used to create more resilient systems.

The Giving Tree

In Shel Silverstein's classic book *The Giving Tree* (1964), a boy is friends with an apple tree. At first, he takes just the apples the tree has to offer him, but as time goes on, he takes more and more from the tree, not just apples. He cuts off pieces of the tree to use for various things until all that's left is a stump. The tree wants nothing more than to please the boy, but in the end, it has nothing left to give except a place to sit. It's a sad story. But each time the boy took something from the tree, the tree was happy that it was serving a new purpose. If only the boy could've inoculated that stump with mushroom spawn! Trees are giving in nature, just like in the book. Especially with mushroom growing, every part of a hardwood tree has a purpose. The small branches at the top can be used as firewood kindling or to make mulch. The larger-diameter branches make great shiitake logs. The trunk can be used in the woodstove and to create oyster mushroom totems. And,

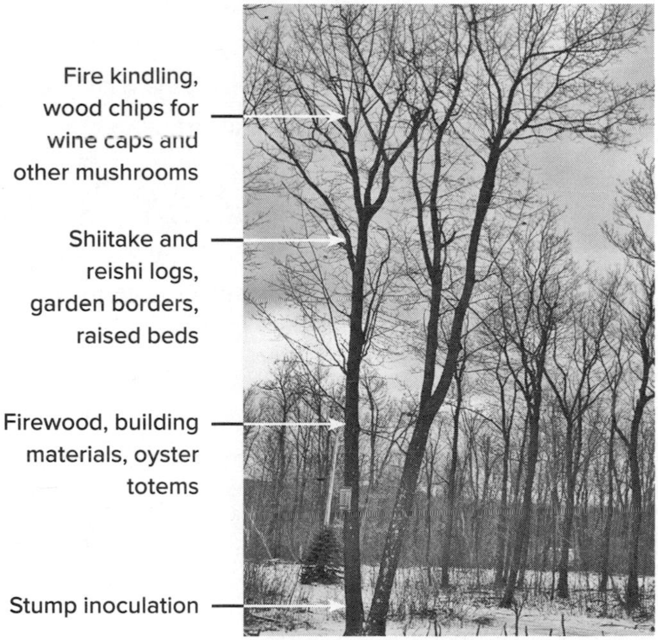

Fire kindling, wood chips for wine caps and other mushrooms

Shiitake and reishi logs, garden borders, raised beds

Firewood, building materials, oyster totems

Stump inoculation

FIGURE 5.4. The Giving Tree.

unlike in *The Giving Tree*, the stump is not just for sitting. It, too, can be inoculated with mushrooms. Every part of the tree serves a purpose. When performing a selective cut, make the most out of each tree, so they are not sad that they have nothing left to give.

Organically Growing Mushrooms

As it grows, a mushroom absorbs almost anything it comes in contact with, through the air, water, and substrate. Therefore, it's best to grow mushrooms using organic methods to ensure that your mushrooms are pure and free of unwanted chemicals or other contaminants. Only inoculate mushrooms onto untreated substrate and situate them where no harmful chemicals will be applied. If your neighbor treats a lawn right next to your property, avoid growing mushrooms in that area. If you are installing an edible landscape, use untreated mulch. Buy organically grown straw for your oyster cultivation, or grow it yourself. Be sure that your rain barrels are not picking up contaminants from the structures the water flows off of. You do not need to be certified organic to use organic methods and to let people know you do. If not certified, you cannot call your market goods "organic," but you can state that you use organic methods.

Thinking Local

A global movement to think local is well underway. From supporting local businesses to eating local foods, thinking local makes sense from a sustainability perspective. Using your local assets, as well as producing and selling locally saves resources, both in terms of money and energy. Less transportation energy is used and less money has to be spent on shipping and transport of goods.

Optimizing Your Property

Creating a resilient landscape design optimizes your property's assets and saves you money. Remember the permaculture principle: "Produce no waste." Instead of buying a fabricated shade structure for your shiitake stacks, place them in the shade of your on-site forest. Inoculate your forest with spawn. Spread mushroom spores in your woodlot where you think they will grow. You can learn how to do this in *Sepp Holzer's Permaculture* (2004). Focus on forest farming techniques. Inoculate your stumps after a selective cut, making those trees the best giving trees around, demonstrating the permaculture principle of using and valuing renewable resources and services. Once you learn more about mushroom cultivation, grow your own spawn to save money and shipping. Though not in the wheelhouse of this introductory book to mushroom growing, there are great resources out there that teach you how to make spawn. We recommend *The Mushroom Cultivator* (Stamets and Chilton, 1983), *Organic Mushroom Farming and Mycoremediation* (Cotter, 2014), and *Growing Gourmet and Medicinal Mushrooms* (Stamets, 2000) for those who want to learn how to produce their own spawn and create indoor growing spaces. See Resources for more.

Sourcing Materials Nearby

Reducing your expenses and helping to conserve natural resources, use materials already available on your property or nearby, whenever possible. If you have hardwood forest on your land, for example, you can cultivate mushrooms in that forest and source materials from the forest to grow your harvest. You can use the natural shade of that forest to place shiitake logs, instead of constructing a shade structure, which uses natural resources and costs money.

Selling Locally

By necessity, purveying the freshest mushrooms requires you to sell them locally, as they have a short shelf life. But, this also saves transportation energy and costs. If you are producing mushroom grow kits, you may wish to start shipping them to customers. To keep your business as environmentally friendly as possible, you may consider not offering expedited shipping, which has a greater energy footprint than standard shipping.

Water Conservation

Mushrooms require a great deal of water. Therefore, as part of your resilient design, you want to minimize the use of water as much as possible. Some strategies for water conservation include:

1 Capture rain water, such as through a rain barrel system.
2 Grow mushrooms near water bodies so that the mushrooms soak up evaporating water and, if grown on the ground, soak up more water from the soil.
3 Grow mushrooms near plants so that they soak up transpiring water and CO_2 from the plants.
4 Grow mushrooms in areas already receiving water, if possible, such as in an existing greenhouse or in your gardens and edible landscapes (don't place mushroom species that are very susceptible to contamination in the greenhouse, i.e., oyster grow bags, but shiitake and other logs, like reishi, are fine to locate there).
5 Water your mushrooms early in the morning or late in the evening to minimize fast evaporation, conserving water.
6 If growing mushrooms indoors, limit traffic in and out of the grow room, not only to minimize contamination, but also to retain your desired humidity, therefore saving water and energy.

FIGURE 5.5. Wine caps with drip irrigation.

7 Apply water directly to your mushrooms and plants, such as through drip irrigation or misting.

8 Consider using timers and automatic shut off systems so that you do not forget to turn the water off and use much more than you need.

Companion Planting (Pairing): A Polyculture Approach

Companion planting, or *pairing*, is the process of coupling species together in mutually beneficial growing situations. They can benefit one another in a variety of ways. Pairing is a demonstration of the permaculture principles "integrate rather than segregate" and "value diversity" because it creates mutually beneficial relationships between a variety of complementary species. Developed by Native Americans, a commonly known example is the "three sisters," in which corn, beans, and squash are grouped together in a planting. In this relationship, corn provides beans with a natural pole to grow upon. Beans fix the soil's nitrogen, improving soil fertility in future years. Squash's expansive leaves prevent weeds from growing under the corn. Each also provides nutrients to the others that they do not possess, rounding out a great example of mutually beneficial relationships. Mushrooms can be a fourth sister, providing benefits that we will discuss shortly. Another example is the addition of native perennials to attract pollinators, serve as habitat for beneficial insects, beautify the garden, and provide shade for wine cap (and other) mushrooms.

Wine Caps and Other Mushrooms in Your Garden

There is a serenity that takes over when you walk into your garden and see veggies, herbs, and mushrooms, all growing a great smorgasbord of healthy, tasty food for your family. A truly healthy garden is also a unique and diverse environment that can be a great classroom for your kids, friends, and family—or even your customers if you hold on-site workshops or invite them to visit. Our garden was home to vegetable plants, mushrooms, native perennials, praying mantids, toads, garter snakes, a diversity of pollinators, and a whole host of nature's creatures. Considering that the garden rested on top of a mountain, the first order of business was to build soil. We do a lot of composting, so grass cuttings were the initial mulch in the early years. Hoeing and tilling were our primary weeding measures until we learned more about incorporating mushrooms and about permaculture methods.

An example of mushrooms in a pairing is the production of wine cap stropharia within a vegetable garden. We paired wine caps and kale in our garden. The early- and fast-growing kale served to shade the wine caps naturally, while the mycelia and unharvested fruiting bodies of the wine cap enriched the garden's soil over time and served to keep the soil moist. The result was a robust and continually producing wine cap patch and large, fortified kale with a superb taste. This was according to our farmer's market customers, who swore that our kale was the best they ever tasted. We even had a few "juicers" who reserved multiple bags of kale ahead of time or lined up for it first thing in the morning, trying to get us to sell them our kale before the sound of the market bell, which was against the rules. We've found that kale enthusiasts are much like mushroom lovers…they're all outlaws. Though most of our wine caps grew under kale, we've found that other great pairing options are zucchini, squash, tomatoes, or under a bean teepee, so, basically under any leafy garden plant that will provide sufficient shade.

FIGURE 5.6. Wine caps under tomato plants.

One of the most interesting phenomena that occurred with our wine cap companion planting was that tomato plants with wine caps underneath did not get infected with the late season blight (*Phytophthora infestans*), after we experienced years of complete failure due to this scourge. Organic gardening and fighting the blight are tough rows to hoe, pun intended. We absolutely hate and despise the blight. Too many times, all of our hard work trying to grow tomatoes with organic methods turned to blackened blobs of mush. The winters of empty shelves where jars and jars of tomatoes should have been were simply heart-wrenching. But then, maybe thanks to wine caps, we got Cherokee Purple, Arkansas Traveler, and Zapotec tomatoes, even when all the rest of our plants were yellow-leafed with blackened fruits.

So, this simple observation became the hopeful focus of an effort to figure out what, if any, role wine caps had in combatting blight. We tried some more unofficial experiments with similar results and then worked with a plant pathologist to develop a scientific study design. Perhaps the aggressive nature of the wine cap mycelia was outcompeting the blight? Or, were the wine cap's spores the heroes? Dave's theory of "my fungus is tougher than you, Mr. Blight," just didn't cut the mustard scientifically. We still do not know if or how wine caps play a role in late season blight suppression. We moved to a semi-arid, blight-free climate before we could implement the study. So, though we do not have blight anymore (yippee kai-yay!), we also cannot study this phenomenon locally. But, we feel it's worth further exploration. So, who's in? If you have late season blight in your area and would like to participate in a multi-site North American study, contact us via mycelialmayhem.com and we will set you up with our design, as long as you agree to share the results with us and be acknowledged for your participation.

Mushrooms in Your Greenhouse

Another pairing strategy is placement of shiitake or reishi logs under greenhouse plants. If you have a greenhouse, use the open space under your benches to grow mushrooms. The space is already shaded and humid. You already water your greenhouse plants, which flows down beneath your planting tables. Why not use this shade and water to grow mushrooms? This saves water, money, time, and potentially the need for non-renewable resources like a shade structure. It also increases your production (and profit, if selling mushrooms) per square foot by utilizing vacant space for maximum yield. The mushroom mycelia produce CO_2, which is needed by your greenhouse plants, so the practice demonstrates multiple permaculture principles.

Composting

Excess mushrooms and mycelium make an excellent addition to your compost pile, as they enrich the resulting soil. Mushrooms, such as blewit (*Clitocybe nuda*) and almond agaricus (*Agaricus subrufescens*) can be grown on compost. This can give you an additional mushroom patch and a richer compost pile.

Recycling Spent Growing Medium

Straw

When we were in full-blown production, we went through bales and bales of straw for oyster production. When you grow oysters on straw you will get two, maybe three or four, flushes, and then they diminish fairly quickly and become host to aggressive molds, other fungi and fungus flies, thus threatening your overall total production. Our simple cure was, after the second or third flush, we would haul the bags out to the garden, cut them open and lay out the straw as mulch in the garden. When we became full time "mushroomers," we didn't have the time to weed our garden like we used to. So, laying out copious amounts of

spent oyster straw sure did make it a lot easier. Even though from a production standpoint the straw was spent, it never failed that, after a rain or a watering, we'd start to see oysters hiding under tomatoes, beans, squash, and every other plant we had in the garden. It is a pleasant surprise to go to gather herbs and veggies and see a nice cluster of oysters to add to your dinner or take to the farmers market. We would often add straw to these areas to squeeze even more oysters out of the patches. With all of that straw and all of those mushrooms, we had the richest, darkest,

FIGURE 5.7. Oysters sprouting in the garden.

deepest soil on the mountain. Our garden was a thing of beauty to behold while sipping on coffee, watching the sunrise, or figuring out which job was on the day's agenda.

Logs

On our first foray growing mushrooms in the garden, old shiitake logs were used to create an "herb wheel" raised bed. We simply placed "spent" five-year-old logs, which were not producing many mushrooms anymore, in a hexagon and then placed more logs in a wagon wheel formation, with a sage plant in the middle. Serving as a hub in our garden, the herb wheel almost always became a point of discussion for visitors. The logs are a wonderful building material and, as they further break down, they enrich the soil. Surprisingly, the herb wheel logs kept producing shiitakes, here and there, well into their tenth year after inoculation.

We also made other uses of shiitake logs. When you are inoculating 200+ logs a year, you will find after about 3–4 years of production, sometimes longer or shorter (depending on strain and species of log), your logs aren't producing that well. We created an asparagus bed out of spent logs, notching them as you would if building a log home. We then filled in with soil and compost and planted asparagus crowns in. Every

spring one of our favorite veggies comes up and sometimes a nice couple of caps will pop too, even with no bark and the logs crumbling. We also created a sister bed to our asparagus bed with a Hugelkultur bed in between (described below). In this bed we grew Jerusalem artichokes, which are a great root vegetable for people with diabetes, and they taste great too. These native sunflower plants would sprout up and provide shade for the shiitake log frame and

FIGURE 5.8. "Spent" shiitake logs sprouting in the garden.

FIGURE 5.9. Shiitake log and oysters.

FIGURE 5.10a. Hugelkultur raised bed creation. Shiitake logs as base.

FIGURE 5.10b. Base covered with straw.

spent oyster straw in the middle. In the fall when everything was winding down, you simply flip the soil, gather the tubers and eat. Remember to leave enough "seed" tubers for next year's crop, though.

We also built a Hugelkultur raised bed using spent shiitake logs. Using methodology from Sepp Holzer (2004) and some of our own ideas, we placed them seven wide and two long on the bottom, five wide and two long on top of those, and so forth, creating a pyramid until we had a structure one log wide on top. We took spent straw and filled in as many of the gaps as possible between the logs. Then, we topped it off with a layer of sod, upside down, and soil on top. This method helped us to grow tomatillos and other heat seeking veggies successfully that we hadn't been able to before. This mounding of spent logs gave us an additional gardening area, and it was vertical, so its production didn't require increasing its footprint.

FIGURE 5.10c. Finished mound.

Materials and Energy Reduction

Successfully capturing, storing, and conserving energy is vital in optimizing your mushroom growing operation and your business, as is materials reduction. Here are some strategies for energy and materials resilience:

1 Use a natural wood fire, rather than propane, for straw pasteurization or find another alternative pasteurization technique.
2 Use more outdoor strains appropriate for your climate and grow your mushrooms in natural environments, minimizing the need for energy intensive pasteurization and indoor growing.
3 Power your grow house with solar energy.
4 Transform existing outbuildings to grow rooms.
5 Use plant-based biodegradable grow bags.

What else could we do to grow with more resilience?

Business Resilience

Building a resilient site helps to make your business resilient by saving you time, money, and natural resources. Also, by optimizing the operation with nature's time-tested principles, you increase your efficiency and success. Do you want your business to survive or thrive? It's just as important to build resilient business systems as it is to create a resilient physical landscape. You can just as easily apply nature and permaculture principles to social systems, like businesses, as you can a landscape. Here are some ideas to get you started, but the sky's the limit.

1 Diversifying your product line, services, and income stream—permaculture principle: "use and value diversity."
2 Diversifying the people involved and engaged in your operation, getting continual feedback, feedforward, creative ideas, different perspectives, and input.
3 Collaborating with the community and other businesses for mutual benefit.
4 Keeping an eye on present and future market conditions, anticipating, innovating, creating, and embracing change, so that your

business can adapt to uncontrollable outside forces. What niches need to be filled now and in the future?

5 Capturing solar energy to reduce your long-term business costs and create resilience.

See chapter 8 for more information on creating a resilient business.

Sharing Your Bounty

Applying biomimicry to social systems is one of Kristin's passions, having initiated a nature-inspired leadership model while attending graduate school. Our greatest joys in mushroom growing are in sharing them with others and sharing our knowledge as well. We loved to mimic a mushroom's spores, spreading the wealth of mushrooms with the wind so that the passion and science of mushroom growing can be replicated and reproduced, eventually taking root and spreading out like mycelium through our growing networks of mushroom enthusiasts. We hope people will grow fruit of their own and share their bounties, perpetuating the cycle to repeat in more communities and ecosystems. Whether in a mushroom-related business or not, how are you sharing your bounty and cycling valuable resources for "the betterment of all life forms?"

THE FRUIT OF YOUR LABOR

Umami of Edible Mushrooms

What Is Umami?

Umami is derived from the Japanese word, うま味. Though difficult to describe, umami translates to "pleasant, savory taste" and was scientifically proven as the fifth taste sense by Kikunae Ikeda (1909), a professor of the Tokyo Imperial University. It is considered the fifth of the five basic tastes, along with sweet, sour, bitter, and salty. If it's not one of those, chances are it's umami. The pleasant, warm taste is often mild, but flavorful, and lingering. Mushrooms, along with other foods rich in glutamate, such as meats, vegetables, and cheeses, are often described as savory or umami foods. A notable exception is the medicinal reishi, which is bitter. Some examples of dishes rich in umami are soups, stir fries, and pasta dishes. Glutamate describes the components of glutamic acid, a protein or amino acid, responsible for the taste. Use this chapter for your own kitchen, and if selling mushrooms, communicate this information to your customers to enrich their experiences with your mushrooms and keep them coming back for more!

Often, mushrooms have an umami taste that is unique to mushrooms or unique to a particular variety of mushrooms. The umami of mushrooms has been described as savory

FIGURE 6.1. Umami means "pleasant, savory taste."

with a woodsy note. However, the unique umami of varieties is quite diverse, from the woodsy shiitake to the lobster-esque lion's mane. In 1957, Akira Kuninaka (1960) realized that the ribonucleotide GMP present in shiitake mushrooms also conferred the umami taste.

From Harvest to Feast:
Handling Mushrooms for the Best Umami

From the grow block or garden to the plate, the proper handling of fresh mushrooms preserves their freshness and therefore, maximum umami. It also ensures food safety, so be especially sure of your handling procedures, i.e., have a written food handling/safety plan if a purveyor. If selling mushrooms, educate your customers on the proper post-sale storage, prep, and cooking methods so that they come back for more. For instance, stress the importance of storing mushrooms in a dry paper bag or specially-made mushroom bags, cooking all mushrooms thoroughly, and trying small portions of new mushrooms to ensure that they are not sensitive to certain varieties.

Harvesting

Timing is everything when harvesting mushrooms, especially to ensure umami. With many mushrooms, umami turns into "eew-mami" once the fruiting body starts to sporulate. Know your varieties of mushrooms well so that you harvest them at the perfect time. For instance, wine cap stropharia should be harvested when the fruits become large enough, but the caps are still "closed" and burgundy in color. Shiitakes will have a well-formed cap that has not turned upward to start sporulation, as is the case with many capped mushrooms. Oysters will often give their own signs of spore development. The clusters will be well-developed, stacked on top of one another, and spread outward prior to sporulation. A *dusting* from the fruiting body above will appear on the clusters once they start producing spores.

Remember to pull your hair back when harvesting to prevent any strands from getting into your mushroom bags. Wear a clean shirt, free of pet hair or other dirt. And, of course, wash your hands prior to harvesting your mushrooms.

To harvest cleanly, follow these instructions:

1 Tie hair back and make sure you have clean clothing on and have washed your hands before you pick the mushrooms.

2 Determine whether the mushroom is fresh. You do not want to pick anything past its prime.

3 Most mushrooms should be pulled, not cut, to increase their storage longevity. While picking, inspect the mushrooms for bugs, which can hide underneath the gills. If present, brush bugs of with mushroom brush. Brush off any pollen or dirt.

4 It's best to weigh (if applicable) and package mushrooms on the spot (in paper bags or produce containers), at the site of harvesting, to avoid contamination from the local environment, unless you have a certified kitchen to work in. An example of potential contamination is the pet hair that floats through your house, including in your kitchen.

5 Label containers with harvest date, and if applicable, price.

6 Transport harvest to a clean refrigerator with a temperature of 45–50°F.

Storing

With fresh mushrooms, proper storage is paramount, both pre- and post-sale. Harvested mushrooms deteriorate quickly, some more than others. While shiitakes and wine caps have a fairly long refrigerator life, oysters do not last that long, and the amount of time varies according to the species or variety of oyster. We've had shiitakes last as long as a week (sometimes even more), while our oysters only last

FIGURE 6.2. Mushroom storage bag.

three or four days. It's a good thing that mushrooms don't sit unused for long in the houses of their enthusiasts. Though mushrooms vary in their longevity, here are the primary rules of fresh mushroom storage:

1 **Refrigeration:**
   a Always store mushrooms in a dry container allowing for some air circulation; it could be one made for produce, a loosely closed paper bag, or a specially-made reusable mushroom bag, which is usually made of breathable cotton and features a drawstring. (If a purveyor, you could make these with your logo imprinted and sell them full of mushrooms for an upcharge.)
   b Keep packaged mushrooms between temperatures of 45–50°F, even in transport.
   c Never store mushrooms in a refrigerator's crisper, as this area is too moist and will cause premature decay. Instead, place your mushrooms on the shelves.
   d Ideally, use mushrooms within a few days.

2 **Freezing:** When freezing mushrooms, you want to take precautions to ensure the best flavor. We like to grind mushrooms up in a blender with a little olive oil and some walnuts (or pine nuts) until it is a thick paste consistency. We put the mixture into ice cube trays, cover with wax paper, and place in the freezer. When solid, we break out them out, wrap individual cubes in wax paper, bag in small quantities and vacuum seal before placing in the freezer again. When we need some for cooking, we pull our cubes out individually. Alternatively, you could cook the mushrooms lightly with some bouillon, cool, and use the ice cube tray process. Having some liquid (water or oil) seems to lock in the flavor best and allows you to have some pre-measured mushrooms on hand, even in the winter. Fresh mushrooms can be chopped up and frozen without any additions, but you will need to vacuum seal them to properly preserve them. Dave's Nana used to cook up a batch of mushrooms with bacon, eggs, and flour and freeze it in old butter tubs (see recipe "Nana's Mush" near end of this chapter). Lightly cooking a mushroom side dish and freezing it works fine.

3 **Dehydrating:** This ageless process is excellent for preserving mushrooms, and for some, it seems to intensify the flavor. A basic food
   dehydrator will do the job. Dry them thoroughly and vacuum seal
   or store in canning jars. Shiitake, black trumpets (chanterelles),
   oysters, etc., can be dehydrated whole or sliced. We slice and
   dry reishi too, though very small ones can be dried whole. All of
   our cooking rubs start with dehydrated mushrooms ground to a
   powder, and are finished with spices and sometimes salt. If you
   are selling mushrooms, remember that a bag of dried mushrooms
   costs a lot more than the same bag of fresh. Also, without the water
   content, they are more concentrated. Let your customers know
   that and keep it in mind when cooking.

4 **Canning:** We have never liked the texture of jarred mushrooms.
   Our family "puts up" jellies, jams, whole tomatoes, tomato sauces,
   pickles, etc., but we've never found any recipes that do mushrooms
   justice. We have had mushrooms that were blanched and run
   through a pressure cooker that were very tasty and had a nice texture. We are still in the process of understanding pressure cooking,
   though, so we don't feel prepared to offer instructions here.

Processing and Preparing

Because *all* mushrooms need to be cooked thoroughly before consumption and because you are careful to consume organically grown
mushrooms from a reputable source, thorough cleaning, in the traditional sense, with lots of water or a solution, is not necessary. In fact,
such cleaning will compromise the integrity of the mushrooms, causing
them to become soggy. To clean mushrooms, wipe with a moist, clean
towel or with a dedicated, soft mushroom brush. You want to eliminate
the dirt and any bugs hiding up under the gills beneath the cap.

Cooking

The number one rule of thumb with mushroom consumption is to
cook all mushrooms thoroughly. If you have ever perused the salad
bar, you have seen raw button mushrooms as a topping. Not only do

FIGURE 6.3. A medley of mushroom ingredients.

these buttons have very poor umami, especially when raw, but they, along with portobellos, criminis, and false morels, can contain hydrazines, chemicals that some scientists believe cause cancer (FC&A Medical Publishing, 2004). Also, uncooked mushrooms will just pass through your system undigested due to their chitin cell walls. Therefore, the nutritional value is lost. Cooking mushrooms is simple and easy. Most can just be sautéed in butter or olive oil with some onions and green pepper for a delicious snack, spread, side, or grilled meat topping. Finally, some people have a sensitivity to mushrooms, so advise first-time eaters to try a small portion of every new mushroom variety before they dive into larger amounts.

Complementary Food Pairings for Umami and Health

Mushrooms, much like other savory foods, have rich umami that is, ironically, a bit like a blank slate that can be paired and seasoned diversely. Foods with various nuances of umami can be paired very successfully together. For instance, mushrooms, with their savory taste pair well with vegetables, baked, grilled, or sautéed meat, salt, and a variety of spices. They are so versatile, you can work up a dish in no time from your garden's varieties and your available mushrooms.

The Taste, Texture, and Presentation of Specific Mushrooms

To assist with your culinary use of gourmet mushrooms, here are Dave's descriptions of how our favorite mushrooms taste, feel, and look:

Shiitake: earthy, woodsy, rich, can have a hint of garlic; meaty texture;

FIGURE 6.4. Let's see, what can we make from these fresh garden ingredients?

can be basis for a dish, such as a superior replacement for portobello sandwiches; long used for its medicinal qualities.

Oyster: most have a mild flavor, but the tastes differ; these are unique and versatile mushrooms, the deeper the color, the richer the flavor (we've found).

Pink Oyster: a lot of people say that it has a mild seafood flavor (we've never tasted that, but we eat lion's mane for its seafood flavor); short shelf life; beautiful pinkish/salmon color, which can be somewhat preserved with very fresh oysters sautéed over low heat.

Italian Oyster: thick caps; rich, woodsy flavor; crunchy texture; supposedly awesome with wild boar, but we've never tried it.

King Oyster: big, meaty, rich mushroomy flavor; great in creamy mushroom soup; use the stem also.

Lion's Mane: seafood-ey, firm texture, sweetish; looks really unique and cool; long used for its medicinal qualities.

Pioppino: the best, great classic mushroom flavor; a bit of a crunchy texture; full of umami.

Chicken of the woods: tastes like frog legs (just kidding), therefore tastes like chicken; firm texture, rich aroma; can be basis for a dish—a great replacement for hormone-infested industrialized chicken.

Maitake: very woodsy, great in soups, great with long grain rice; usually, specimens are big enough to serve as base for a couple of dishes; long used for its medicinal qualities.

Black Chanterelles: earthy, great complement as a dehydrated seasoning, awesome in crème (i.e., our cream of watercress and black chanterelle soup, see recipe to follow); presentation mushroom, as they are small and shiny black.

Nameko: slimier than eel skin until you cook it, then it's awesome; nutty, crunchy.

Wine Cap Stropharia: meaty, mild, woodsy flavor, great steak topping, great cooked with wine and sherry, which bring out its sweet side; can use entire mushroom (cap and stem), so great presentation, especially because of the burgundy cap.

Morels: there's a reason they're the most sought after!

Reishi: tastes like sucking on a dirty sock (worse than bitter), learn to add honey, lemon, ginger, flavored teas (like raspberry), or a mixture of spices or spiced tea; long used for its medicinal qualities.

Culinary Supplements

Because a lot of people do not care for the texture or consistency of fresh mushrooms, we produce "culinary supplements," which contain various mixtures of dried, powdered mushrooms and spices. We market

FIGURE 6.5. Our chicken rub.

the mixtures by providing recipes and dish ideas for each packet. To customers who do not like fresh mushrooms, we explain that, by using the supplements, they receive the benefits of added flavor and nutrients without having to deal with the texture of fresh mushrooms…the umami without the "eew-mami." These supplements can be as versatile as you want, considering that endless combinations of mushrooms and spices can be concocted. These mixtures can be used as rubs, in soups, or as dip mixes; many can be used as all of the above.

Recipes

The following recipes are presented by Dave.

ENTREES & SIDES

Mushroom Stuffed Venison Medallions

1 lb venison "Butterfly Steaks" or other lean game, bison or beef into ½" thick 2"–3" rounds with cut almost all the way through

¼–½ lb fresh mushrooms (stropharia, shiitake, pioppino)

1 medium onion, diced

6 oz fresh spinach

6 oz shredded provolone

2 cloves garlic, crushed

1 tsp salt or to taste

½ tsp black pepper or to taste

Italian bread crumbs

toothpicks

Heat a skillet to medium heat, sauté mushrooms, onions, and garlic until onions are translucent, set aside. Tenderize steaks with a meat tenderizer until thin. Lay out steaks and layer on top sautéed mushrooms, spinach, and cheese, sprinkle with salt and pepper to taste. Roll up and place toothpicks to hold together, dust in Italian bread crumbs. Heat skillet to medium-high heat, fry 3 minutes per side, pat dry and serve.

Grilled Duck with Berry Mushroom Sauce

6 mallard duck breasts (other
　　similar sized ducks can be
　　used)
salt and pepper
olive oil
3 cloves minced garlic
Sauce:
1 cup red raspberries
½ lb fresh mushrooms (shiitake,
　　pioppino, wine cap stropharia,
　　and oyster all work)
1 medium onion, minced
½ cup dry red wine
2 tbsp butter
1 tbsp honey
1 tsp thyme finely chopped
coarse ground pepper

Duck is one of the hardest game animals to cook. You want the breast meat at a rare to medium-rare consistency, while still rendering the fat. The trick is to cook quickly over high and cook slightly longer on the skin side. Rub duck with olive oil, garlic cloves, and salt and pepper. With a sharp knife, make 6–12 holes with the point of the knife through the skin but not too deep. Over hot coals place the duck breast skin side down, approximately 3 minutes for rare and up to 6–7 minutes for medium-rare. Turn over with tongs (forks will poke the meat, which will dry it out). Place on a plate, skin side up, and let sit for 5–7 minutes. Slice thinly across the grain and pour sauce over. Garnish with remaining raspberries and serve with a side salad and wild rice side dish.

　　Sauce: In a sauté pan, melt butter over medium heat, add onions and mushrooms. When onions start to become soft, add red wine, thyme, and approximately ½–¾ of the berries. Leave the rest for garnish. Simmer for 8–10 minutes and slowly add honey, stirring to thicken. Add a few cranks of a pepper mill, to taste. Pour sauce over duck after it has set for a while.

Mussels and Shrimp with Oyster Mushrooms over Linguini

½ lb of shrimp
½ lb of mussels
1 lb of oyster mushrooms
1 tbsp butter
2 tomatoes, diced
4 green onions, chopped
1 tbsp chopped garlic
5 fresh basil leaves, chopped
capers, to taste
1 red pepper, chopped
1 tbsp lemon juice
1 tbsp white wine (Pinot Grigio
　　is a good choice)
1 box linguini

Boil linguini. Heat skillet to medium heat. Add mussels and shrimp. Cook down approximately 5 minutes. Add butter, garlic, onions, and mushrooms. Cook until onions start to become translucent. Add red pepper, wine, basil, tomato, and lemon juice. Cook an additional 3–4 minutes. Add capers. Cut off heat. Drain linguini and top with mushroom and seafood mixture.

FIGURE 6.6. Mussels and shrimp with oyster mushrooms over linguini.

Holiday Mushroom Stuffing

8 oz bread crumbs
½ pint oysters
6–7 chestnuts
2 celery stalks
½ c fresh oyster mushrooms
　　(or shiitake)
1 onion
1 c chicken broth
sage, salt, pepper

Chop chestnuts and oysters, then dice onions, celery, and mushrooms. Mix all ingredients in a bowl, while slowly adding chicken broth until you reach your desired consistency. Stuff mixture in bird and bake at 350°F for 1 hour per pound. For larger birds, like turkey and goose, double the ingredients. This stuffing accompanies our Christmas Eve goose (or duck, pheasant, or wild turkey), and we have also used it for Thanksgiving. (Our mushroom-loathing family said it was the best stuffing they've had in a long time, not realizing that the superior umami came from the oyster mushrooms. Dave is so sneaky and bad sometimes!)

Mushroom-Rubbed Standing Rib Roast: QED—Quick, Easy, and Delicious

rib roast
2 tbsp dehydrated mushrooms
　　(we prefer stronger-flavored
　　mushrooms for this recipe)
¼–½ lb chopped fresh
　　mushrooms, preferably same
　　varieties as dehydrated
2 garlic cloves, minced
1 medium onion, minced
1 tbsp Worcestershire sauce
1 tbsp red wine
½ c beef broth
salt and pepper

Bring roast to room temperature, set oven on broil. Rub dehydrated mushrooms, salt, and pepper on roast. Place roast on baking sheet or sturdy foil, fatty side up. Broil until fat begins to crisp. Bring roast out of oven and place into a roaster, rib side down and fatty side up. Bring the oven down to 375°F. Add minced onion and garlic, Worcestershire, wine, broth, and fresh mushrooms. Cover with foil. Place roast back into oven for 1 hour. Turn off oven. Leave roast in oven, *but do not open oven door for 3 hours*. About 30 to 40 minutes before serving time, turn oven to 375°F and reheat the roast. Important: Do not remove roast or re-open the oven door until ready to serve.

Mushroom-Stuffed Fish

4–5 whole trout, scaled, head attached (a small whole walleye could be substituted)

½ c (or more, depending on size of fish) fresh mushrooms sliced (morels, shiitake, oysters, or chanterelles)

2 slices of bread broken into pieces

1 tbsp fresh parsley

1 tbsp lemon juice

1 medium onion

2 tbsp butter

2 cloves garlic, minced

2 tbsp teriyaki sauce

salt and pepper to taste

If using charcoal or grilling over wood, start fire first. On stove, sauté mushrooms, onion, garlic, and parsley in butter. When onions become translucent, stir in lemon juice and bread. Cut off heat and let cool until you can handle with ease. Cut a piece of aluminum foil slightly longer than the fish and lay the fish on top of the foil. Fill the cavity with the mushroom stuffing and brush a light coating of teriyaki over the fish. Add salt and pepper, if desired. Cover with extra foil and wrap tightly, twisting the ends if necessary. Lay fish on hot grill, approx. 4–6" above coals, if using charcoal or wood. Flip fish approx. 3–4 minutes later. Smaller fish take less time than bigger fish, so be careful not to overcook. A 4–6 lb walleye or trout will take about 5–8 minutes per side, while a small brook trout only requires 2–3 minutes per side. If your guests are squeamish, simply place the finished fish on a plate and remove the head with a knife, and serve. Or, if you have a cleaver, pay tribute to *A Christmas Story*.

Red Beans and Rice with Shrimp

½ lb shrimp, cut into ½" chunks

½ lb oyster mushrooms

2 tbsp vegetable oil

1 medium onion, chopped

1 medium red pepper, chopped

2 stalks celery, chopped

2 tsp salt

1 tsp freshly ground black pepper

5 cloves garlic, minced

3 bay leaves

1 tsp dried thyme

1 tsp hot sauce

½ tsp cayenne pepper

1 lb red kidney beans, rinsed and picked of debris

water, to cover mixture

1 c rice of your choice

Soak beans overnight in 2 quarts of water, picking out any debris. Simmer beans until they are soft (not mushy!). Place the vegetable oil in a medium saucepan and set over medium-high heat. Add mushrooms, onion, bell pepper, celery, salt, and pepper to the pot. Cook, stirring frequently, until the onions and celery are semi-translucent and the bell pepper is tender, 6 to 8 minutes. Add the garlic and cook for 1 to 2 minutes, stirring constantly. Add bay leaves, thyme, hot sauce, cayenne pepper, water. Add semi-cooked beans and increase the heat to high. Cook, stirring frequently, until the mixture comes to a boil, approximately 6 to 8 minutes. Decrease the heat to maintain a simmer, cover and cook for 1½ hours, stirring every 30 minutes. Uncover and increase the heat slightly to maintain a steady simmer and continue to cook for another 30 to 40 minutes or until the beans are tender and the sauce is thickened to your liking. Add the shrimp when you have 15 minutes left. Prepare rice during the last 30 minutes of cooking the beans, according to package. Serve the beans over the rice.

Mountain Man Meatloaf

2 lbs ground venison, bison, or
    lean beef
6–8 slices bread, your choice
    (cut into small pieces)
1 lb shiitake mushrooms,
    chopped into pieces
1 onion, finely chopped
2 carrots, shredded
2 stalks celery, chopped
1 packet dried onion soup mix
3 eggs
1 c beef bouillon

Combine meat, bread, onion, carrot, celery, onion mix, eggs, and shiitake mushrooms in a bowl. Mix thoroughly. Form into loaf. Pour ½ c bouillon over, enough to coat the top and a light layer on bottom. Cover with foil and bake at 350°F for 45 minutes–1 hour. Uncover and strain off excess liquid. Add remaining bouillon for au jus or to make gravy. Place meatloaf back into oven on middle rack and broil for 5–15 minutes, watching to make sure it doesn't burn. Serve with mashed potatoes. Serves 5–7.

Mushroom Omelet

¼ lb fresh mushrooms
1 small onion
1 garlic clove
butter
2 eggs
1 tomato, chopped
1 or 2 cooked bacon strips,
    crumbled (or crumbled
    sausage)
¼ c grated cheese

In a medium saucepan, sauté mushrooms, onions, and garlic in butter, about 10 minutes over medium heat. Put aside. Beat eggs. Heat pan to medium and add a pat of butter. When melted, pour in beaten eggs and add mushrooms. Stir and reduce heat for a few minutes. Add bacon, diced tomatoes, and grated cheese quickly. Cover when the majority of the cheese has melted. Fold in half. Cook for a minute or two and flip, cooking for two more minutes or until all cheese is melted. Serve with a bagel or English muffin, and have fork ready to fight off any poachers!

Mushroom Casserole

2 c fresh bread crumbs
Parmesan cheese, grated,
    to taste
1 c light cream or milk
1 lb fresh mushrooms
butter
marjoram or thyme
1 tbsp onions, chopped

Remove stems from mushrooms and layer in a buttered casserole dish. Sauté butter and onion, combine with bread crumbs, and place over mushroom caps in casserole. Sprinkle with grated cheese and marjoram or thyme. Pour the cream over all. Bake for 30 minutes in a 350°F oven. Serve immediately.

Sautéed Chicken of the Woods

3 c chicken of the woods
(*Laetiporus sulphureus*)
mushrooms, cleaned
1 tbsp olive oil
3 cloves garlic, minced
2 c tomato sauce
½ c dry white wine
salt and pepper to taste

Clean mushrooms with a damp cloth, and then either tear or chop them into small pieces. Warm the olive oil over medium heat and add the garlic. Let it cook for one minute. Add the mushrooms and cook for 10 minutes, stirring occasionally as they turn a vibrant orange. Pour in the white wine and cook for another 5 minutes. Add the tomato sauce and let the whole thing simmer for another 10–15 minutes. Great as a side or tossed with risotto!

Nana's Mush

1 lb mushrooms, ground or diced
very small
1 onion, medium
2 eggs
2 tbsp flour
4 slices bacon
salt and pepper, to taste

Fry bacon, drain. Add mushrooms and onion, sauté over low heat until onions translucent. Add eggs, salt, pepper, and flour until thickened. Can be frozen in quantities for a winter side dish.

DIPS

Creamy Mushroom Dip

4 tbsp dehydrated black
chanterelles
1 block of Philly cream cheese
3 tbsp fresh chives, chopped
1 tbsp French onion dip

Place the chanterelles in a mortar and pestle. Grind until fairly fine and dusty. Blend all ingredients until thoroughly mixed and allow to sit for about ½ hour. Serve with pita chips or crackers. You can also top with fresh slice cucumber, tomato, or lox.

Shiitake Dip

¼ lb fresh shiitakes
½ c walnuts or pine nuts
3 tbsp Parmesan cheese
3 tbsp chopped chives or garlic
scapes
olive oil
lemon juice (optional)

Sauté shiitake in olive oil over medium heat for about 5 minutes, stirring constantly. Mix all ingredients in a blender or food processor, except lemon juice. Blend until mixed thoroughly. If you want a creamier dip, slowly add additional olive oil. Add a few drops of lemon juice at the end and blend. This dip can be served cold or warm. Put in the middle of a plate surrounded by baguette slices and watch it disappear!

SOUPS

Cream of Watercress and Black Chanterelle Soup

32 oz chicken broth

2 qt heavy cream

½ lb fresh chanterelles

1 tbsp dehydrated chanterelles

1 lb watercress

2 tbsp butter

1 bacon slice

1 medium onion, chopped

2 tbsp fresh garlic, minced

2 tbsp dry sherry

½ tbsp thyme

In a pot, brown the slice of bacon without overcooking. Remove bacon, leaving the drippings. Add the butter. Sauté onions and garlic until the onions get translucent. Add the broth, watercress, dried chanterelles, and thyme. With a hand blender, blend the broth until the cress is chopped fairly fine. Bring to a slow boil and then turn down heat. Add fresh chanterelles and sherry. Simmer for 10 minutes. Stir in cream and simmer. Top off with a sprig of watercress. Allow guests to add salt and pepper to taste.

Hot and Sour Soup

12 c water

chicken bouillon

1 lb shiitake mushrooms

1 c Asian greens, chopped

½ c spinach, chopped

1 onion

1 lb tofu, strips or cubes

2 eggs

1 c brown sugar

¼ c Srirachi Hot Chili Sauce

3 or 4 dried hot chilies (serrano, cayenne, or other favorite spicy pepper)

Bring water to a boil, adding enough bouillon for 12 cups of broth. Drop eggs slowly into boiling water. Add chili sauce and brown sugar to taste. Add remaining ingredients and simmer for ½ hour. Serve over rice or with chow mein noodles. Serves 5–7.

SAUCES AND GRAVIES

Sauce de Crème de Chanterelle Noire

2–3 c fresh black chanterelle
    mushrooms
¾ c dry white wine
1 pint heavy cream
2 tbsp butter
2 cloves garlic, crushed
1 tsp salt or to taste
½ tsp black pepper or to taste
1–2 tbsp flour

Lightly sauté butter and garlic in a medium saucepan until flavors mingle. Add white wine and cream. Bring to a simmer and reduce heat to medium-low, stirring frequently. In a medium saucepan, dry-sauté the chanterelles until excess water has evaporated and transfer to the cream mixture. Add salt and black pepper. Simmer on low heat for 25–30 minutes until the sauce has been thoroughly infused with the flavor of the chanterelles. Add 1 to 2 tbsp of flour to thicken.

White Oyster Sauce

1 cluster of oyster mushrooms,
    between ½–1 lb
⅓ c extra virgin olive oil
1 medium onion, chopped
6 garlic cloves, finely chopped
½ tsp dried hot red pepper flakes
¼ tsp dried oregano
⅓ c dry white wine
⅓ c flat leaf parsley, chopped

Heat oil over medium, then sauté onion and oysters (rip up the clusters into small individual petals), stirring, until onions are translucent, about 4 minutes. Add garlic, red pepper flakes, and oregano. Cook, stirring occasionally, until garlic is golden, about 2 minutes. Add wine and simmer, uncovered, stirring occasionally, until slightly reduced, about 3 minutes. Serve over pasta. Optional: Add 8 oz of chopped clams, shrimp, or scallops and sauté slightly before adding onions and oyster mushrooms. Add parsley as topping.

Stropharia Steak Sauce

¼–½ lb wine cap stropharia
    buttons and stems
3 tbsp olive oil
1 small onion, sliced
2 cloves garlic, chopped
⅓ c red wine (Cabernet,
    Burgundy, or other rich-
    bodied red)
1 tbsp Worcestershire sauce
    (optional)

Clean wine cap buttons and stems. Slice ⅛"–¼" thick length-wise. Heat olive oil in pan over medium. Add mushrooms and sauté until golden. Add garlic and onions, flip and continue to sauté until golden brown on both sides. Lower heat to medium-low. Add wine (and Worcestershire, optional). Simmer for 5–6 minutes. Pour over steaks fresh off the grill.

SUPPLEMENTS & RUBS

For each of these rub recipes, start with ½ cup dehydrated mushrooms, blend and/or grind; add about 1 part spices for each 3 parts mushrooms, to taste. Mix it up for different tastes and according to which mushrooms you have available to you.

Mycelial Mayhem Chicken Rub

Chives, parsley, sea salt, black peppercorns, and chicken of the woods, chanterelles, and a blend of oyster mushrooms. Good on fish too!

Italian Oyster Chicken Rub

Oregano, chives, thyme, flat leaf parsley, sea salt, black peppercorns, and a hint of oyster mushrooms.

Oyster Lamb Rub

Oyster mushrooms, chives, thyme, sage, flake salt, black peppercorns, and a dash of apple mint.

Gravy and Soup Amender

Chicken of the woods, black and golden chanterelle mushrooms, combined with shiitake, oyster, lion's mane, and pioppino mushrooms with flake salt, and black peppercorns! Add to gravy or soups for a rich woodsy flavor!

Italian Mushroom Rub

Italian oyster, pioppino mushrooms, Greek oregano, Genovese basil, garlic, and onions. Add to any Mediterranean cuisine, great on steak too!

Pork's Pal

Oyster mushroom, sage, thyme, sea salt, chives, a hot pepper, and a sweet pepper. A great rub on pork. Also good with a dry white wine poured over fish.

Steak Rub

Shiitake mushrooms, rosemary, chives, onion, flake salt, black peppercorns. Rub on steaks, or marinate in red wine and pour over steaks. Great in/on burgers too.

Spicy Steak Rub

Shiitake mushrooms, rosemary, onion, flake salt, black peppercorns, and a hot pepper of your choice, according to your heat preference.

APPETIZERS

Duxelles

4 tbsp unsalted butter
3 c finely chopped shiitakes
5 shallots, minced (or green onion)
½ tsp coarse salt
¼ tsp freshly ground pepper

In skillet, melt butter over medium-high heat. Add mushrooms, shallots, salt, and pepper and cook, stirring frequently for 10 to 15 minutes or until liquid has evaporated. Can be refrigerated for 10 days. Use to season soups, stews, and sauces or as a topping for vegetables.

*You can freeze this for a great mushroom addition to any dish. Try adding ¼ c of chopped oily nuts (walnuts, cashews, almonds), then pour into ice cube tray. Wrap your "cube trays" in wax paper, vacuum seal, and keep in freezer until needed.

Duxelles: The Appetizer

1 c duxelles (from above)
½ c Parmesan cheese
⅓ c sour cream
pinch of thyme
baguette sliced into ½-inch toast points

Preheat broiler. In bowl, mix duxelles, Parmesan, and sour cream. Spread evenly on bread slices; arrange on baking sheet, and broil until hot and bubbly. Serves 6–8.

Huba Pomoc (Slovak for "Mushroom Helper")

1 tbsp dried shiitake powder
¼ c walnuts
¼ c garlic scapes (or nodding onion)
1 tsp cooking sherry
1 tbsp butter
olive oil

In a blender or food processor, finely chop walnuts and scapes (or nodding onion). Add 2 tbsp olive oil. Heat saucepan over low. Coat bottom of pan with olive oil. Sauté mix from blender and add cooking sherry and shiitake powder. Add additional olive oil slowly until mixture is coated. Bring back to sauté and turn off heat. Add butter until melted and mix. Serve with sliced baguette or another artisan bread.

TEA

Reishi Tea

Add two 3-inch antlers or one 4-inch conch of dried reishi to 32 oz of water, simmer for 2 hours. Add ginger slices or lemon, 3–4 flavored teabags and honey to counter the bitter taste. You can reuse the reishi for a second, milder batch.

Nutritional Benefits of Mushrooms

We are inundated every day. Whether we're watching our favorite sport or TV show, it always happens: some super pill shows up during commercials. "Big pharma" can cure everything...funky toenails, floppy wieners, and bald heads. You name it, there's a pill for it. We sometimes enjoy a morbid chuckle when they list the possible side effects: internal bleeding, elevated blood pressure, stroke, and even death. Even death!! Really? Do people need a head full of hair that badly? In our narcissistic society, we will do anything for vanity, even if it kills us. It's sobering that big pharma can pay for so many ads. Apparently, a sick society is really good business; a healthy one, not so much. Is it not better to be healthy? Well, that requires discipline and desire. So why work at it when you can just pop a pill? Because there's a better way, one that benefits people and society.

For millennia, people have been consuming mushrooms for their nutritional qualities. Without the benefit of modern nutritional science, people ate mushrooms to nourish themselves, even despite the risk of accidentally eating a poisonous one. In the wild, modern-day animals do not typically consume poisonous plants or mushrooms; they eat species that their bodies find appealing and need the most. We'd like to argue that early humans had similarly enhanced food instincts. Without the "sense" of modern nutritional science, their "sense" of food nutrition was probably enhanced, much like the keen hearing of a blind person. At the very least, they learned over time which mushrooms to avoid and which ones tasted good, passing this information down from generation to generation. Why stress this? Because, as a mushroom lover, you might hear from "fungi-phobes," mushroom haters, that the nutritional benefit of mushrooms is unknown, unproven, or unclear. We say "Nonsense!" Mushrooms are intuitively, and empirically super-foods.

In Kisaku Mori's 1974 book, *Mushrooms as Health Foods*, he recounts how Wu Shui, a Ming dynasty (1368–1644) doctor talked about the health benefits of shiitake in terms of stamina, health preservation, cold curing, circulation benefits, and cholesterol lowering. Wu Shui did not have the scientific data we have today, yet was right on target with regard to shiitake benefits. Despite rich time-tested traditions of mushroom nutrition and a growing body of scientific knowledge, there's still a lack of mainstream knowledge about myco-nutrition, especially where Western medicine is concerned. You'll still see a lot of naysayer "information" out there, claiming that not enough data exists to support some of the superb nutritional and medicinal benefits of mushrooms. As Western medicine focuses more on treatment than prevention and cure, the thought processes of people within a Western culture have, in the past, stifled the intuitive senses that point to prevention and even the focus on related scientific research. But, fortunately, that's changing.

It's important to:

1 Use your gut, your instincts, to believe that mushrooms have a multitude of nutritional benefits (this is *not* meant as an instruction to go out and *use your sniffer* to safely harvest wild mushrooms… just to trust that you are correct to invest in, grow, consume, and market safely grown or harvested mushrooms for their benefits).
2 Know the current facts about mushrooms' nutritional benefits.
3 Know where to continue to look for new information.
4 Know how to use nutritional facts to market your mushrooms, if applicable.

Nutrition and Medicine

What is the difference, if any, between nutritional and medicinal benefits? From an Eastern medicine point of view, not much. Food is nutrition. Nutrition is the best medicine (via prevention). Therefore, good, wholesome, nutrient-rich, pure (contaminant-free) food is the ultimate medicine. From a Western medicine perspective, "medicinal" implies treatment or cure of some illness that already exists. Sure, nutrition is still associated with prevention, but a disconnect exists between nutrition and medicine. As people in Western cultures are educated about Eastern medicinal philosophies and they start to realize that treatment is big business, this disconnect is fading. But treatment over prevention still has a stranglehold on Western medicine, likely due to the massive quantities of profits made when people are sick and remain so. What if reishi, as part of a healing diet, exercise, and meditation (just an example) could eventually replace or even do better than destructive chemotherapy in treating cancer? Will we ever find out or use it widely if so many have so much money to lose with the death of chemo?

It is also a problem that we tend to compartmentalize health issues, as if our bodies were not systems with interdependent, interconnected parts that all function as a whole. Even some of the most health-conscious of us tend to do this. If you want to prevent heart disease, you focus on consuming heart-healthy foods and take resveratrol supplements, which have their benefits, of course. But, realize that your body is a system, just as an ecosystem in nature. Yes, every part plays a role, performs a unique function, but everything is interconnected *and* interdependent—the immune, the nervous, the circulatory, the skeletal, the digestive systems, and even the skin. Though, by definition, we compartmentalize in this book because we're speaking specifically of mushrooms, keep in mind that mushrooms are part of the equation, as are all other superfoods, exercise, stress management, and even spiritual well-being. So, share the boiled down information in this chapter with friends and family, and, if selling mushrooms, with your customers.

Learning from the Docs ————————

In the 1990s, Dave heard Dr. Andrew Weil speak at the Mushroom Festival in Telluride, Colorado. After raving about the bands he heard echoing off the canyon walls, Dr. Weil got serious and talked about the healing power of mushrooms, natural medicine, and healthy lifestyles. Dave was inspired and appreciated how Dr. Weil explained the complex in simple ways, calling for people to get "back into the kitchen and combat the trend toward processed and fast food."

Dave went back to Pennsylvania and talked to his dad, a family physician, about medicinal mushrooms. Dr. Michael Sewak was intrigued by the information. Though trained in Western medicine, "Doc" understood the human body as a system and was always open to alternatives, like acupuncture. He prescribed nutrition, prevention, and natural methods whenever he could. (Long before Kristin and Dave knew each other, Kristin's entire family went to Doc. Once, her grandmother was in the hospital with heart issues. When Kristin was visiting, Grandma Rosie looked down at her hospital food with a face of disgust that only she could muster. She handed Kristin some money and said, "Go get me a cheeseburger, I can't eat this garbage." Kristin replied, "But isn't Doc making his rounds soon? You'll be in big trouble if he catches you." Rosie had it all figured out, reminding Kristin that "Doc can be heard coming from a mile away. He whistles, he hums, and he jingles his pocket change!")

As Dave talked to Doc about mushrooms, he learned that Doc frequently prescribed natural treatments and that, as Doc would say, "each person needs an individualized diagnosis and treatment, according to his family history, ethnic background, lifestyle, occupation, and even personality." Doc knew each patient's family history, sometimes generations back. Also, Doc explained that the brain is the most powerful medicine or detriment.

He once gave this prescription to a patient: "Eat a bowl of chicken soup, take a shot of whiskey, go around the table three times clockwise and then three times counterclockwise. You should be ok in the morning." Dave shook his head, but Doc explained, "It worked for her, as I suspected it would." Doc also remembered the woman—clinically healthy and ready to be released from the hospital—who told Doc that she'd be dead by midnight. She was right. Doc explained to Dave that mushrooms and other nutrition work to keep many healthy, but not all. If the patient doesn't believe in their powers, then the mushrooms won't help them, or they'll refuse to take them altogether. The man not only understood the human body, but also the human psyche.

Like a mushroom, Dave soaked up all of this knowledge from Dr. Weil and from Doc, as well as from many other sources since his transformational experience at the Mushroom Festival. He realized that, not only was his father much cooler than most doctors of Western medicine, but also that individuals have to be treated as individuals and work with a special doctor to build a health plan that works for them. For us, that includes lots of mushrooms. What can we, as proponents of medicinal mushrooms learn from Doc, from Dr. Weil, and from many others with wisdom to share?

Nutritional Content of Mushrooms

As part of a well-balanced diet of mostly fruits, vegetables, lean protein, and complex carbohydrates, mushrooms can play a significant role in a person's good health. Though almost all water, mushrooms are good or excellent sources of protein, fiber, and complex (healthy) carbohydrates. Plus, they are typically high in vitamin D, riboflavin (B_2), and niacin (B_3), while also containing other B vitamins and some calcium. They are also high in important minerals like potassium, selenium, and copper. Finally, mushrooms are gluten-free, vegan, low-calorie, cholesterol-free, fat-free, and low in sodium but high in "umami" (see Chapter 6), which can reduce a person's need for meat and salt.

Of course, nutritional content can vary greatly from one mushroom to another. So, getting to know which mushrooms offer the greatest or a targeted benefit can inform your selection of mushrooms to cultivate and consume. Of the edible culinary mushrooms, shiitake and maitake stand out as the top medicinals because of their diversity of benefits. Other edibles have great benefits as well, and a host of other "non-culinary" mushrooms are excellent medicinals, which are used in supplements and infusions.

Mushrooms are what we call a *fundamental* vegetable, which we consider to be the veggies that should comprise a large part of our daily diets due to their high nutritional content and minimal sugar or carbohydrate content. People can eat as many mushrooms and other fundamental vegetables as they want, even if on a low carb (complex only) diet to lose weight, control diabetes, cure insulin resistance, or prevent cancer. It's important to note, however, that even though mushrooms are in the "unlimited" fundamental category, they should be slowly incorporated into your diet if you have not eaten a lot of mushrooms in the past or are trying a new variety. Fundamental vegetables are "unlimited" because they are very low calorie and low carb, plus they pack superior nutrition.

Fundamental vegetables include:

- Mushrooms
- Artichokes
- Asparagus
- Bamboo shoots
- Bean sprouts
- Broccoli and cauliflower
- Cabbage
- Carrots
- Celery
- Cucumbers
- Eggplant
- Garlic
- Green beans
- Greens and kale
- Okra
- Onion, chives, leeks
- Peppers
- Radishes
- Sea vegetables
- Tomatoes
- Yellow, summer and spaghetti squash
- Zucchini

The Known Benefits

Eating the right medicinal mushrooms, in the right quantities, as part of a healthy, balanced diet, carries a myriad of known health benefits. According to Wasser (2010, Abstract), "a total of 126 medicinal functions are thought to be produced by medicinal mushrooms and fungi, including anti-tumor, immunomodulating, antioxidant, radical scavenging, cardiovascular, antihypercholesterolemia, antiviral, antibacterial, antiparasitic, antifungal, detoxification, hepatoprotective, and antidiabetic effects."

Say what?

With so many functions, let's boil the massive amount of information that's out there down to the basics, the primary benefits of mushrooms. In essence, mushrooms serve so many medicinal functions—regulating, modulating, optimizing, and protecting your system—that, perhaps more than any other food, they contribute to whole body health. They are superfoods, and listed as such in many of the health education resources out there. But, it's not just about the mushrooms. So, when consumed with a wide variety of other superfoods, and in the absence (for the most part) of harmful foods, such as processed white sugar, you can optimize your health and greatly increase your chances for a long and happy life. As an example, we've compiled a list of top anti-cancer foods:

- Medicinal Mushrooms: Reishi, shiitake, maitake, lion's mane, etc.
- Berries
- Chlorella and Spirulina
- Cruciferous Vegetables: Turnip greens, broccoli, cabbage, cauliflower, kale, etc.
- Fish
- Garlic
- Ginger
- Grapes
- Green Tea
- Hemp
- Oats and Barley
- Sea Vegetables: Wakame and kombu
- Tomatoes
- Turmeric

The *Super Seven* Medicinal Mushrooms

Though research continues to add to the body of knowledge, we know a lot about how mushrooms contribute to our health. Mushroom guru Paul Stamets participates in and summarizes much of the modern scientific literature on the medicinal properties of mushrooms. From his summary of medicinal properties of certain mushroom species (2005), seven mushrooms top the list for their diversity (number of different) of health benefits. We've ranked them according to the number of different health benefits provided by each. Reishi tops the list, with 16 major benefits, with cordyceps, shiitake, and maitake taking the two, three, and four spots. Chaga, turkey tail, and zhu ling tie for the rank of five, each possessing seven of the 17 listed effects.

TABLE 7.1. *Super Seven* Medicinal Mushrooms.

Rank within Group	Super Seven Medicinal Mushrooms	Antibacterial	Antitumor	Immune Enhancer	Antiviral	Liver Tonic	Blood Sugar	Stress Reducer	Blood Pressure	Kidney Tonic	Lungs/Respiratory	Cholestorol Reducer	Anti-Candida	Nervous System	Antioxidant	Anti-inflammatory	Cardiovascular	Sexual Potentiator
1	Reishi or Ling Chi (*Ganoderma lucidum*)	X	X	X	X	X	X	X	X	X	X	X	X	X	X	X	X	
2	Cordyceps or Caterpillar Fungus (*Cordyceps sinensis*)	X	X	X	X	X	X	X	X	X	X	X		X	X		X	X
3	Shiitake (*Lentinula edodes*)	X	X	X	X	X	X	X	X	X	X		X	X				X
4	Maitake or Hen of the Woods (*Grifola frondosa*)	X	X	X	X		X	X	X			X		X	X			
5	Chaga (*Inonotus obliquus*)	X	X	X	X	X	X									X		
5	Turkey Tail or Yun Zhi (*Trametes versicolor*)	X	X	X	X	X				X					X			
5	Zhu Ling (*Polyporus umbellatus*)	X	X	X	X	X					X					X		

Adapted from Stamets (2005)

Mushrooms' Major Health Improvement Functions

By no means is the following list exhaustive, but here are some of the major ways that mushrooms protect your immune and other systems, help to treat ailments, and have the potential to reduce or cure disease and other unfavorable conditions:

1 **Control Blood Pressure.** Dave knows, intimately, that mushrooms help to control blood pressure. A few years ago, he went to a doctor and found out that he had borderline high blood pressure. Though the doctor proposed a prescription (synthetic) medication, Dave decided, after going home and thinking about it, to try a reishi regimen first. After a few months, Dave's BP was in the normal range, and he hadn't changed his diet otherwise. In addition to reishi, shiitake, cordyceps, and maitake work to lower blood pressure, according to Stamets (2005). Shiitake and maitake are great fresh edibles, while reishi and cordyceps are normally consumed as a supplement or as an ingredient in tea or other infusion.

2 **Lower Cholesterol.** Reishi, cordyceps, and shiitake have been found to lower cholesterol. Reishi works through steroid-like compounds to inhibit cholesterol development (Stamets, 2005), while shiitake reduces cholesterol through an amino acid, which helps the body more quickly metabolize ingested cholesterol and dispose more of it as waste (Mori, 1974).

3 **Kill Bacteria.** All of the mushrooms we call the *Super Seven* (see chart) work as natural antibacterial agents, and other mushrooms do as well. Unlike taking antibiotics, mushrooms work as natural antibacterial agents, and along with other nutritional elements, work to keep our systems in balance and prevent bacterial illnesses, such as digestive flora imbalances.

4 **Strengthen the Immune System.** Through a host of functions, such as immunomodulation, anti-inflammatory, antioxidant, and

FIGURE 7.1. Reishi tea being made at the Sewaks.

free radical scavenging, mushrooms help to regulate and protect immune function.

5 **Kick Cancer's Ass.** How? Let us count the ways: through detoxification, anti-tumor, immunomodulating, antioxidant, antiviral, anti-inflammatory, blood sugar regulation, alkalizing, and radical scavenging actions. You already know that antioxidants prevent free radicals from forming. And, you probably already know that inflammation can lead to cancer and other unwanted conditions.

But, you might not know the myriad of ways mushrooms could play a role in preventing, treating, or even curing (at some point) certain cancers. How does a mushroom's role in regulating blood sugar matter in cancer prevention? Well, cancer feeds on sugar. So, if too much sugar is regularly present in your system, your cancer risk is greater. How does a mushroom's alkalizing effect prevent cancer? Well, in addition to feeding on sugar, cancer prefers a low body pH. What does a mushroom's antiviral properties have to do with cancer prevention? Well, some cancers are initiated by viruses.

This is only the beginning of our exploration into the potential of mushrooms, both as prevention and treatment of cancer and other illnesses. Could mushrooms, somehow, help your body break down that protective barrier that some cancer cells have, allowing the immune system to identify the invasion and respond more quickly? Someone should research that potential because of the myriad of known anti-cancer functions of medicinal mushrooms, such as immunomodulation (moderating or achieving the proper immune response) and anti-tumor effects. And, who knows what mushrooms can do in conjunction with other natural nutritional and medicinal techniques? The research of treatment through healing foods and disease-curing lifestyles is exploding! Here are some of the ways mushrooms prevent and could treat cancer:

a **Prevention!** With over 100 health functions, mushrooms are a superfood, performing as part of a healthy overall diet and lifestyle to prevent disease, including cancer. Mushrooms help to prevent cancer through antioxidant, antiviral, radical scavenging, anti-inflammatory, blood sugar regulation, body-alkalizing function, and many others. An ounce of prevention is worth a ton of cure.

b **Action against Specific Cancers.** For brevity's sake, let's look at the *Super Seven* only, as examples of specific mushrooms acting against specific cancers, according to Stamets (2005). He identifies specific cancers that six out of the *Super Seven*, plus a few other mushrooms, combat.

TABLE 7.2. Mushroom Activity against Specific Cancers.

Rank within Group	Mushroom Activity Against Specific Cancers	Breast	Cervical/Uterine	Colorectal	Gastric/Stomach	Leukemia	Liver	Lung	Lymphoma	Melanoma	Prostate	Sarcoma
1	Turkey Tail or Yun Zhi (*Trametes versicolor*)	X	X		X	X	X	X			X	
2	Maitake or Hen of the Woods (*Grifola frondosa*)	X		X	X	X	X				X	
3	Reishi or Ling Chi (*Ganoderma lucidum*)					X	X	X			X	X
4	Shiitake (*Lentinula edodes*)	X					X			X	X	
5	Cordyceps or Caterpillar Fungus (*Cordyceps sinensis*)					X		X	X			
6	Chaga (*Inonotus obliquus*)		X									

Adapted from Stamets (2005)

c **Anti-Tumor.** One of the most amazing and exciting areas of medicinal mushroom research is their potential ability to stop tumor growth and therefore, possibly metastasis (spread of cancer cells to other tissues), or even to shrink tumors. Why is this so exciting? Ask anyone who's gone through radiation or chemotherapy and they will describe the devastating impacts both can have on your body (they attack your healthy cells, too). Also, some medical professionals suspect that chemotherapy has the potential to promote future cancers in other areas of the body by damaging healthy cells and therefore leaving them susceptible to cancer. All of the *Super Seven* mushrooms have anti-tumor activity, some of which, like reishi, have great immunotherapy potential because they are not toxic in high doses (Stamets, 2005).

d **Chemotherapy Coping.** Mushrooms may help people cope with chemotherapy symptoms through their anti-inflammatory, immune enhancement, and other healing properties.

6 **Strengthen Bones.** According to FC&A Publishing (2004), mushrooms are the *only* non-animal whole food source of vitamin D for your body! This is a *huge* fact for all vegetarians out there, for postmenopausal women with osteoporosis, and people with lactose intolerance. Your body can only get vitamin D from the following sources:

a Internal production (which is not enough).

b The sun (which can be dangerous in excess).

c Animal foods that contain vitamin D (which, if consumed in excess, can cause problems).

d Vitamin supplements (which could serve you well, but experts are finding that getting your nutrition from whole foods, rather than extracts, is more effective).

e Mushrooms, especially shiitakes, chanterelles, and maitake.

7 **Stabilize Blood Sugar.** Of the *Super Seven* mushrooms, the top five help with blood sugar: Reishi, cordyceps, shiitake, maitake, and chaga. Therefore, mushrooms help with diabetes, cardiovascular problems, and even cancer.

8 **Protect the Brain and Nervous System.** Lion's mane (*Hericium erinaceus*) not only can look like a brain, but it has been shown to protect the brain and larger nervous system. According to Stamets (2005), lion's mane contains nerve growth stimulants, which promote neuron and myelin tissue regrowth. This could help to treat a wide range of nervous system problems, such as Alzheimer's, muscular

dystrophy, and neurological trauma. It could even increase cognitive ability.

9 **Alkalize the Body.** Though on the low end of the alkalizers, mushrooms have a mild alkalizing effect on the body, which is a good thing. An alkaline body is usually a healthy body because our cells function best when at a slightly alkaline pH. An acidic body, on the other hand, can produce numerous problems, from allergies to cancer, and many more issues in between. To alkalize your body, consume mushrooms and the much more powerful alkalizers, like spinach, broccoli, celery, green beans, zucchini, and garlic. See a pattern developing? Eat those low-carb veggies! Limit foods with an acid-producing impact on your body, such as refined sugars, beef, pork, wine, beer, and coffee. Are you thinking what we're thinking, "Crap, I can't have any of the good stuff!" Well, we can at least try to tip the scales in our favor, right?

10 **Increase in Stamina and Recovery from Fatigue.** According to Mori (1974), shiitake mushrooms help people recover from chronic fatigue and increase stamina, including sexual stamina. These two related effects are likely due to shiitake's ability to lower cholesterol, improve circulation, and balance hormones due to their vitamin and mineral content. Removal of cholesterol from the blood, with enhanced circulation, translates to healthy body cells free of toxins, which means more energy, therefore, an increase in stamina and less tiredness.

11 **Antiviral, Antifungal, and Antiparasitic.** According to Stamets (2005), certain mushrooms hold promise in treating AIDS and other viral ailments because of their strong antiviral activity. Also, because cancer is sometimes caused by viruses, mushrooms' antiviral function plays yet another role in fighting cancer. Candida, a fungal parasite of the gut, is very harmful to the digestive system, but can be stamped out by reishi, maitake, lion's mane, and shiitake (Stamets, 2005). However, colon therapist Brenda Watson (2002) recommends avoiding mushrooms when trying to eliminate Candida. These contradictory recommendations demonstrate the importance

of working with a natural health care professional, such as a naturo-pathic or integrative doctor when trying to prevent or manage specific conditions.

And about 100 additional benefits are known (Wasser, 2010).

How to Maximize Nutritional Content
Grow Organically

Mushrooms are now being cultivated for mycorestoration of polluted areas because of their excellent ability to soak up everything, including toxins, around them. (To learn more about mycorestoration and other breakthrough uses of mushrooms, read *Mycelium Running* by Stamets [2005].) But when grown for consumption, you want to control what your mushrooms can soak up. Grow all of your mushrooms using organic methods and substrates. What's the point of eating mushrooms for your health if their benefits are negated by the toxins contained within?

Companion Planting and Quality Growing Media

A simple example of how a mushroom's health, and therefore its nutritional value, can be maximized is companion planting mushrooms with plants that provide sufficient shade. We used the shade of our kale patch to optimize the growth of our wine cap stropharia mushrooms, which were situated below, in a hardwood mulch and straw mixture. Are there any other benefits to the wine cap that is situated atop a rich humus-like layer of garden soil that's been built up and provided nutrients by the plants that have grown there over the years? Perhaps. A very important factor in the nutritional quality of your mushrooms is having quality growing medium.

Food Pairings

We believe that, at a very basic, chemical level, foods that taste good together are also foods that should be paired together for nutrition, in addition to their complementary tastes. A healthy diet consists of lots of fresh veggies, most of which are paired very well together, such as through a salad, mixed in pasta sauce, or in an omelet. A healthy diet also incorporates protein and a small amount of complex, unprocessed carbohydrates, both of which pair well with veggies and one another. The same goes for dairy, unless there's a specific reason why you are to avoid dairy. Focus on whole, fresh veggies, consuming at least two cups daily (of course, including mushrooms!) and the other foods that taste good with veggies, and you'll have a healthier diet that helps to prevent and even treat disease and unfavorable conditions.

Servings and Supplements

Hands down, the best way to get the most nutrition, and therefore medicinal effects, of mushrooms is to eat fresh mushrooms regularly as part of a healthy, vegetable-rich diet. According to Stamets (2005), the USDA identifies a single serving of button mushrooms to be about 84 grams, fresh, or about 8 grams, dried. Though dried mushroom supplements can be used to add nutrition to your diet, they might not contain the amount of mushrooms necessary to make a significant difference in your daily intake. As an example, one supplement Kristin took contained 100 mg of shiitake and 100 mg of maitake. If we assume, for simplicity's sake, that a single serving of each of these mushrooms is nearly 100 grams, fresh, and 10 grams, dried, then this supplement only contains 1/10th of a gram of each, amounting to 1/100th of a daily serving.

This does not mean that mushroom supplements are worthless, as you do not need to consume a whole serving of mushrooms daily to derive benefits. The medicinal effects of many mushrooms do not require whole servings. However, be aware that you will get more benefit from consuming a couple of servings of fresh, super-nutritious mushrooms per week. You can accomplish this by simply incorporating shiitakes into your diet, adding others as you can, and figuring out what, if any, supplement you'd also like to add. Supplements can be a good way to consume the medicinals that are not good culinary mushrooms. But, how can you know how much to take or which variety to choose?

Working with Your Doctor

Because no recommended daily value exists for supplements, and thus there are none for the super medicinal mushrooms, you should consider working with a family physician who specializes in prevention and integrative medicine or a naturopathic doctor to identify what your nutritional needs are in relation to fresh and dried mushroom consumption. Each person's health plan should be different, based on a host of factors (see Dave's Tangents, earlier in this chapter), so find someone who understands this, works with the whole person, and is open to or specializes in alternatives to Western medicine's "just pop a pill" approach. If you have a specific health problem or you'd like to prevent disease, they can "prescribe" the right mushrooms in the right amounts. You may find that you need massive amounts of certain supplements, which is just another great reason to learn how to grow your own mushrooms and eat more fresh ones.

Cautions

"It's his excessive consumption of mushrooms! They've addled his brain and yellowed his teeth!" says Saruman of Radagast in *The Hobbit* (2012). Most of the time, fungi-phobia is unfounded. (They certainly don't yellow your teeth!) But, as they say, there's a bit of truth behind every myth. Some cautions for mushroom consumption are listed here:

1 Button, portobello, and false morels are known to contain hydrazines, chemicals believed to cause cancer (FC&A Publishing, 2004); plus they're not the most nutritious, so it's best to avoid consuming them in favor of others.

2 Wild collection can be dangerous, so know what you're doing or go with an expert.

3 Mushrooms are very absorptive, so only eat organically grown mushrooms.

4 Go slow at first—avoiding a "shiit-take," as each mushroom is different in its composition, some affecting digestion in people who are new to those mushrooms. Also, because its cell walls are made

of chitin, the same complex carbohydrate in shellfish and insect exoskeletons, digestion can, at first, be challenging.

Once, we gave a friend a huge bag of shiitakes as a thank you gift. The next day, Dave ran into him and he didn't look so good. He said, "No wonder they call them shIIt-takes!" Dave asked him how much he ate and found out that he consumed about three meals-worth of shiitakes, and he wasn't used to eating them! So remember to take care when consuming a new type of mushroom.

If you are not accustomed to consuming mushrooms on a regular basis, go slow, eating only a little bit the first time to avoid a shiit-take! If selling your mushrooms to people new to mushrooms or certain varieties, advise them to do the same, perhaps by handing out a general information sheet or card with each sale. People can have an intolerance to all mushrooms or only to certain kinds. Rarely, people can be allergic to mushrooms. Even people without any kind of intolerance should incorporate mushrooms slowly, so that their digestive systems can get used to processing them.

For further information on the nutritional or medicinal properties of mushrooms, see *Fungal Pharmacy* by Robert Rogers, *Mycelium Running* by Paul Stamets, and the other relevant books listed in our Resources section.

Important note: When informing customers of mushrooms' benefits, know when and how you need to protect yourself, such as with a disclaimer: "The information provided here is for educational purposes only and is not intended to diagnose, treat or prescribe for any condition."

Niche Business or Supplemental Income

Do you absolutely love mushrooms? Do they inspire you for all that they do for humanity and because of their beauty? Do you deeply believe that mushrooms can change lives and the planet? If so, you might be cut out for mushroom farming. This is the attitude, the passion, the deep connection you need to start a mushroom venture. But, it's not easy.

> Some people say you can't make a living farming. I just tell them that doing anything else isn't really living at all. —POST ON FARMON.COM (2015)

Mushrooms As a Business?

Do you...

- *Love* doing the same thing over and over again, assembly line style...i.e., inoculation?
- *Revel* in long hours?
- *Enjoy* waiting for up to a couple of years for a harvest?
- *Easily shrug off* insults, such as when a farmers market customer spots your hard-earned shiitakes and yells "Eewwwe, mushrooms" loud enough for all potential customers to hear?
- *Not mind* having to close down your entire grow house because of contamination?

If you answered yes to most or all of these questions, then a mushroom business is for *you*!

As Jody Venn (our brother-in-law, friend, and fellow mushroom grower) of Tsoma Farms in Gainesville, Florida, so eloquently puts it, "Run away while you still can! But if you are in—be in for the long haul; do your customers right and they will do likewise. Along with teaching, agriculture, legal and whatnot [it's] some of the hardest earned bucks—period." Growing mushrooms as a business is *hard*. So many things can go wrong, from the development of the dreaded green mold (or other, nearly limitless possible infestations) to a weak market. And there's more: the low hourly wage (at least at the beginning), the fragile profit margin (especially at the beginning and if you cannot secure great deals on supplies), early farmers markets, and short product shelf life. During his presentations, Dave shows a picture of Brad Pitt with the caption, "Me: Before Mushroom Farming" and then one of himself, with the caption, "Me: After Mushroom Farming."

If you think we're trying to discourage you, we're not. We just want you to know what you're getting into. The primary rule of mushroom cultivation as a business is that you have to have the *desire, commitment,*

FIGURE 8.1. Basket of mushrooms.

and *perseverance* to succeed. Otherwise, stick to hobby growing until you have massive amounts of these qualities. To be a successful cultivated mushroom purveyor, you must go all in, doing your best to maximize production, minimize risk, enhance your skills and knowledge, increase profit margins, and generally working extremely hard. You have to *own it* in more ways than just on your LLC business registration.

You're still in? Great!

Knowing How to Grow

You could just jump in and hit the ground running. But, if you've never grown mushrooms before, you're in for a surprise. You'll experience failure and face trial and error to find what works for you and your particular circumstances, such as climate and how much time you have available. Learning how to grow and what works for you can take a long time, often much longer than the estimated five years it takes to bring a small agricultural operation into "the black." Therefore, we recommend that most people learn how to grow first, before making the decision to start a mushroom-based business. Before David started our mushroom farm, he had been wild collecting for about 40 years and cultivating for more than ten years. See Chapter 3 for detailed growing instructions to get you started, as well as our Resources for in-depth cultivation reference books and other valuable resources.

Formulating Your Idea

"I want to sell mushrooms" is not a meaningful enough idea. What mushrooms do you want to sell and why? What mushrooms are you capable of starting with? What mushroom-derived products, such as DIY grow kits, are you going to sell and why? What other considerations should be made when formulating your idea? For example, does producing dehydrated mushroom products require additional, potentially cost-prohibitive, measures that you're not aware of?

Keep in mind these new business principles, too: Success usually comes either from a new idea in an existing market or from a variation on an old idea in a new market. For example, success could come

FIGURE 8.2. Your unique niche.

from something as simple as the production of fresh, locally grown gourmet mushrooms where they are not available (an old idea in a new market). Alternatively, your idea may be to develop a new mushroom product and distribute it widely within existing markets— a more complex undertaking. Either way, or with both options, you are looking for your own unique business niche. You can find ideas online and either bring them to a new market or transform them for an existing market. If you're creative, you can also formulate a completely new idea. See the figure shown here for a graphical representation of the kinds of niches that translate to business success.

Below are some questions to help you advance your idea of growing mushrooms to more specific plans. Note: Your idea *will* evolve as you develop the rest of your business plan, test your ideas with potential customers, make adjustments, and complete market research. Also important: You will need to keep your answers in mind when formulating your specific goals later in business planning.

1 Why mushrooms? What motivates you to grow and sell mushrooms? You must be in it for the right reasons, and your primary reason should not be profit or you will likely be severely disappointed.

2 What are your values? Sustainability? Organic methods? What defines you? Develop your venture's values statement.

3 Summarize your idea(s) in 50 words or less. If you are planning to focus solely or primarily on mushrooms, you will need to consider incorporating a diversity of products and services to earn enough income to be successful (fresh mushrooms, other mushroom products, a book, recipe books, speaking engagements, on-farm workshops, consulting services, etc.). Product and service diversity, because of the importance of income diversification, is especially important in this tough business. Here's a sample income diversity projection chart:

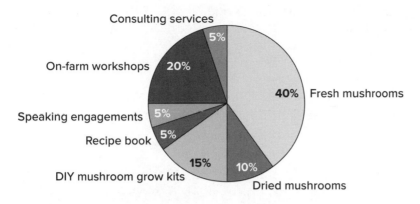

FIGURE 8.3. Product and service diversity.

4 How did your idea(s) originate and where does it fit within the market? Identify each item as qualifying for success under the three niche categories: new idea/existing market; old idea/new market; transformation of old idea/existing or new markets.

5 How are you qualified or uniquely positioned to successfully launch your idea(s)? (Experience, cash investment, education, partners, etc.)

6 In general terms, what does success look like for you? ($_____ in supplemental income; independence by _____ year; profits of $_____ by _____ year, etc.) When answering these questions, it helps to also think in terms of reducing your personal expenses so that you do not need significant profits. What debt can you reduce or eliminate? What purchases can you go without, etc.? How much more of your own food can you grow? Many sustainable farming operations live simply and purposefully from their core values, but also to save money.

7 Generally, what challenges do you anticipate and how do you plan to overcome them?

Once you've formulated your idea, consider obtaining input from outside sources to ensure that it is sound. Don't be afraid to ask for information, advice, or comments. Following are some of the types of people or groups we've found helpful:

1 Agricultural and mushroom trade associations.
2 Your local mushroom club or mycological society.
3 Other growers willing to help you.
4 Government entities such as your state or provincial agriculture department, local extension service, or federal agency.
5 Mushroom lovers—test your fresh product with potential consumers.

Evaluating Your Market: Conducting a Market Analysis

While experimenting with cultivation, complete an evaluation of your market conditions to inform your decision of whether and how to grow mushrooms for profit. Evaluation does not have to be expensive or complex, but you do need to do a market analysis.

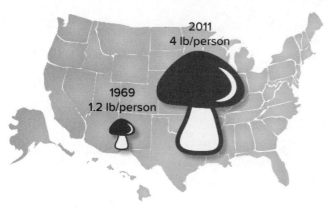

FIGURE 8.4. US citizens' mushroom market. Adapted from USDA ERS (2013).

Mushroom Market Information

Finally, some good news! According to the USDA Economic Research Service (ERS), 2013, people in the United States were eating four times the mushrooms in 2011 than they were in 1969. We made the little infographic shown here to illustrate this phenomenon. If this trend continues at the current rate, demand will rise by one pound per person each decade.

From our experience, and that of other growers, demand is often higher than a small producer's capacity to provide mushrooms, even in a market that is considered on the weak side. Especially in the early years of your mushroom business, you may find it extremely difficult to keep up with demand. We often ran out of mushrooms less than half way through our Saturday farmers markets, and our mushroom market was not as strong as some others. Especially if you focus on finding your ideal customers, you should not experience much trouble selling your mushrooms, unless an unusual density of producers exists locally or local demand is unusually low.

Market-Specific Questions You Need to Answer

Each market is unique. So, if you are primarily focused locally, you will need information about your target areas. Here are some pre-business planning questions you might need to answer.

1 Are there adequate numbers of potential customers in your target geographic area to support your product? Is demand high, and if so, how will you keep pace with demand…or at least remain only a few steps behind? What segments of the population are potential customers? People with gluten intolerance, diabetes, or high cholesterol, those at risk for cancer, vegetarians and vegans, health-conscious omnivores, etc.? How do you find them *and* connect in meaningful ways with them?

2 Who are your potential competitors?

3 What are your competitors' products, prices, services, and distribution tactics? What geographic area do they cover? How do you rate their quality and marketing power?

   a A few pointers on retail product pricing: Start at the grocery store, which, especially in the case of gourmet mushrooms, will have a sub-par product, compared with your fresher, homegrown mushroom. To start, calculate a price that's 20–35 percent higher, and see where you are. There ARE exceptions to the viability of charging a bit more than the grocery store. We've seen, in different areas of the country, fresh shiitakes selling anywhere from $10 to $25 per pound. This is likely due to highly variable climates, with some of the more arid regions requiring distant suppliers and higher transportation costs. So, be aware of where you are and what factors might play into local pricing. Then, visit your local health food store. Your pricing could be about the same as this outlet. Factor in your quality and your circumstances when pricing your product. It is beneficial to take the time to track and roughly figure out your profit margin by knowing how much in materials and time is needed to produce your product. If you determine that $7 in materials and time is needed to produce a pound of shiitakes, and you charge $12 per pound, your retail profit margin is $5 per pound.

b Wholesale product pricing and consignment: When you sell mushrooms wholesale, such as to a local food distribution company, a health food store, or even, in some cases, restaurants, you will receive a wholesale price. In most cases, wholesale is somewhere in the vicinity of half of retail or a bit more. So, using the example above, you would receive $6 per pound for shiitakes, so, you'd be losing money. You need to be sure your retail rate is high enough so that you can still make money when wholesaling. Another possibility is consigning your product to a seller (a store, farmers market seller, co-op, etc.). You could receive a bit higher of a percentage with consignment because the seller isn't buying your product at wholesale prices. Their risk is lower, which can be appealing to a seller, and you don't have to put a lot of effort into selling, which can be appealing to you, but neither of you make any money if the product doesn't sell.

c Determining your own salary: This is less straightforward than product pricing. At the beginning, you probably won't be a salaried employee of your LLC, but you can determine a rate you will be paid (this can be called "consulting" or "billable services" when you are charging outside entities for services). There's no exact formula for this. But you can use and adapt the formula in Figure 8.5 to determine what salary you can afford to pay yourself. Just ask yourself what you would like to be paid per year and divide that by 2080 (the standard number of business or office hours in a year) to calculate your base rate. Then figure in your administrative and general overhead.

TABLE 8.5. General fee for service rate formula.

Person Performing Services	Base Rate	AOE (20% example)	GOE Bill Rate	Revised Base Rate	Benefit Rate	Total Bill Rate	Total Billable Hours	Total Fee Due
Formulas	A	B	C	A+B+C=D	E	D+E=F	G	FXG = H
John Doe	$25	$5	$5	$35	$6	$41	20	$820

Base Rate = Salary + Employer's Share of Social Security (0.062) + Employer's Medicare Cost (0.045) + Employer's Worker's Compensation Cost

AOE = Administrative Overhead Expense = Fundraising, Accounting, Reporting, Auditing, etc. Can change formula above to reflect your AOE.

GOE = General Overhead Expense = Rent, Phone, Internet, Office Supplies, Insurance, Legal, Subscriptions, etc.

Benefit Rate = Retirement, health insurance, and other benefits.

Notes

Total fee does not include additional costs directly associated with performed services, such as travel.

Pre-approval of additional direct costs can be outlined in each service agreement.

4 Is there room for more? Are your competitors struggling to keep up with demand?

5 How will your mushrooms be different from your competition's mushrooms? Does this matter, or is demand so high that difference is not that important?

Getting Started

Structure, Registration, and Insurance

We found that forming an LLC, Limited Liability Company, was the best option for our mushroom business. In most cases with a family farming operation in the US, an LLC is the best option because it limits the family's liability for business-related obligations or debts, protecting personal assets. It also allows owners to report business profits and losses on their personal tax returns, thereby enabling owners to benefit from business-related deductible expenses and qualifying for other tax benefits. As long as you do not need to issue stock to investors or split profit and loss with the business for a lower overall tax rate, an LLC structure will serve your purpose well. Forming an LLC is easy, inexpensive and painless. LLCs also require much less maintenance paperwork than corporations. Registration can be accomplished through your secretary of state or other appropriate state agency.

Important note: An LLC is a "limited liability" company, so it only protects its members from normal business obligations, such as a bank loan. However, if an LLC member has co-signed, personally, on an LLC bank loan, the member is now personally responsible for that debt. So be aware of these aspects if you plan to obtain traditional financing for your business. An LLC does *not* protect its members from all liability, such as professional misconduct, negligence, or personal harm to another. So you will need professional indemnity liability insurance, which should be included in your general business liability plan. For most small farming businesses, a $1 million policy is sufficient. However, consult with a lawyer and an insurance company to determine what insurance you need.

Because LLCs do not exist in Canada, forming a corporation is the best option. You can register your business federally or provincially. If you plan to do business in only one province at the beginning, you should register provincially, with the option of registering later with other provinces within which you plan to do business. Provincial registration is easier than federal registration.

What's in a name? Berglorbeer Farma

We named our farm Berglorbeer Farma. Because of our German and Slovak heritage and our desire to get back to our roots with farming, the name held plenty of meaning for us. Berglorbeer means "mountain laurel" while Farma is "farm" in Slovak. So, translated, our farm was Mountain Laurel Farm, which was appropriate because it was situated in the Laurel Highlands Mountains in the Central Appalachians. Our farm name checked two of the three boxes when it comes to naming a business. It was unique and it had meaning. We didn't have trouble getting a domain name, that's for sure! But it could have been simpler; it's hard to pronounce or understand…lesson learned after having to explain it multiple times at busy events.

Steps to forming an LLC in the US:

1 Pick a company name. For branding and marketing purposes, choose a name that is: (a) unique; (b) simple; and (c) meaningful.
2 File your articles of incorporation with the appropriate state agency.
3 If your state requires an operating agreement, file this as well.
4 In most cases, you will need to obtain an EIN, employer identification number, from the IRS for tax filing purposes.
5 Obtain all industry-specific licenses and permits that are required of your business.

Compliance and Licensing

For agricultural operations in the US, you may have to license your company or specific activities with the appropriate state entity, such as your department of agriculture. For instance, some states may require, now or in the future, the following licenses or permits related to a mushroom farming business:

1 Registration with your state agriculture department.
2 Use of a certified kitchen and correct labeling for food processed prior to sales (in most cases, this includes slicing, canning, drying, freezing, and packaging).
3 A wild collection permit if collecting and/or selling wild mushrooms from public lands.
4 A wild mushroom commercial sales license. (Though this is not currently required in most states, rules, regulations, licensing, and permitting do apply in some states, and new ones are being added in many states at a fast pace to address wild mushroom safety concerns. Federal lands already require a commercial use permit, and activity is restricted on some. See Chapter 4.)

Developing Your Business Plan

1 First, include your overarching idea(s) and niche market information from the previous question lists.
2 What are your business's SMART goals for the next five years?
3 How much money do you need to start and maintain your business?

4 How are you going to finance your initial investment? (Bank loan, community investment such as "crowdfunding," personal investment, etc.)

5 Specifically, what obstacles and challenges do you anticipate, considering your goals and financial plan? How do you plan to deal with them?

6 Specifically, what uniquely positions you for success, and how will you maximize or take advantage of these factors?
   a. In terms of your qualities or assets?
   b. In terms of outside opportunities or market conditions?

7 What resources are available to assist you with your business? (Government, other growers, trade associations, valuable trade shows, farmers markets, etc.)

What the heck are SMART goals, you ask? Those that are Specific, Measurable, Attainable, Relevant, and Time bound. Running each of your goals through these filters will ensure that you can reach your goals in time to make an impact on your business, that they are realistic but important, and that you can measure your success in achieving your goals. Each time you make a goal, make sure it meets these criteria, and you will be setting yourself up for success.

Smart Goals
Specific
Measurable
Attainable
Relevant
Time Bound

Operations

What software or subscriptions will you need to operate your business? Systems and items you will probably need include:
- Checking account
- Accounting software or bookkeeper
- Domain hosting and website development platform, preferably with business email
- Email marketing platform
- Social media accounts, and possibly software that helps you manage them more easily
- Calendar platforms, both for your website and for your internal planning

- Credit card readers, such as those that can process payments through your smart phone or tablet
- Lines of credit with your vendors
- Office equipment and supplies: computers, printer, paper, etc.
- Farming equipment and supplies
- Marketing and event supplies, like a tent, banner, tables and chairs, a display board, and cash box
- Ability to take food stamps, through registration with your state or provincial department of agriculture IF you decide to accept them (We did not participate in this because of the high cost of gourmet mushrooms, and therefore their low popularity with food stamp holders.)

Many more examples exist, from memberships to the appropriate as-sociations (who can often provide your business with various mem-bership benefits) to the use of nearly limitless resources out there that could help you run your business and market your wares. Seek out what you need, and you will find it. Then, ask yourself, "What is my budget for all of this?" Usually, you can start with a free version of a system, with limited functionality and versatility, and go from there, upgrading as you grow.

Cleanliness and Food Safety

Be sure you know how to properly handle your mushrooms, from har-vest to sale. Your agriculture department or extension service will have information regarding food safety and may offer templates for the de-velopment of your food safety and handling plan. Once you have your procedures in line, it wouldn't hurt to communicate these to your cus-tomers (your website is a good place to do this). Mushroom food safety is very similar to that of vegetables, with a few differences.

- If applicable, pull your hair back to prevent any strands from get-ting into your mushroom bags. Wear a clean shirt, free of pet hair or other dirt.
- Wash your hands.

- Do not "wash" your mushrooms. If they have specks of pollen or insects on them, lightly brush everything off with a clean, soft, dedicated mushroom brush.
- It's best to weigh and package mushrooms on the spot (in paper bag or in produce containers), at the site of harvesting, to avoid contamination from the local environment, unless you have a certified kitchen to work in. An example of potential contamination is the pet hair that floats through your house, including in your kitchen.
- Mark the harvest date on the packaging. You can do this easily with a marking pen, labels, or an inexpensive stamp.
- Place packaged mushrooms in a clean refrigerator until sale, or in your clean farmers market cooler if harvesting the morning of market. To avoid getting the mushrooms wet in your cooler, use reusable ice blocks and separate them from the mushrooms with cardboard.
- Make sure you know the shelf life for your mushrooms. A good rule of thumb is to sell mushrooms within two days of harvest. If they do not sell in that timeframe, eat, dehydrate, or compost them.

Human Resources

Unless you have significant investors, you probably won't be able to pay yourself, let alone any employees, for quite some time from startup. Therefore, it's important to think creatively here. Enlist family and friends to help with farming tasks. Also, consider developing a farm apprenticeship program for college students pursuing agriculture as a trade or those majoring in business or marketing. If done correctly, the exchange of labor for knowledge and experience can be a fair trade. You will need to put time into this, as you should mentor and develop your apprentices, not just use them for the free labor. You should also feed them when they are working for you, give them other perks like mushroom grow kits, passes to events, or even a portion of profits if they help out at an event, if you can swing it. A lot of time will also be spent advertising the positions at agriculture departments and in selecting your seasonal apprentices. But the time can be well worth the effort in terms of the relationships you develop, benefit to your farm, and the experience gained by the apprentice. We've included our apprenticeship job description in the Appendix, which you might want to use as the basis for your own.

Teams and Performance

Just because your business is family-owned doesn't mean you should yell, "You son-of-a-#$%*^!" when someone screws up. Would you do that to a co-worker out in the traditional workplace? Probably not. Maximizing your own effectiveness and efficiency, as well as that of your team will pay dividends in terms of product produced and profits made. Even if (actually, *especially if*) your team consists of you and your spouse, you will need to spend time communicating, planning, dividing up roles and tasks, and motivating, in a positive way, one another to perform well, offer creative ideas and input, and solve problems. Sometimes, family members more readily treat one another badly, take each other for granted, fail to communicate, blame each other, and develop other behaviors that they would never exhibit in the workplace. All this can be destructive to family businesses and the family. Just because your business is family-owned does not mean that you can ignore the creation of a positive team environment and the importance of teamwork, leadership, understanding, planning, and communication. It's important to treat your family business as a business. As a small example, don't skip the weekly check-in, which is an opportunity to prioritize and plan the week's work schedule and talk about upcoming events, challenges, finances, and more.

Boiled down and simplified for a family business, here are some key principles and steps you could utilize to make your family business more successful through well-coordinated teamwork and other strategies:

- Just as you need to separate your personal and business finances, keep your business and personal roles apart.
- Identify your business roles and responsibilities based on individual strengths. Much more than skills alone, the strengths of a

person consist of skill + enjoyment. Match roles and responsibilities with individual strengths, as much as possible. Then, match with skills…there will likely be tasks that no one wants to do that still need to be done. You don't need to write out lengthy, perfect job descriptions, but you should write down what your roles and responsibilities are in some way, at least in a short, bulleted list. Who is responsible for…?:

- Leadership
- Finances
- Marketing
- Mushroom Cultivation
- Markets and Events
- Customer Service and Relations
- HR and Internship Programs
- Compliance, Taxes, Permits, etc.

- Agree how you are going to communicate, plan, and help one another. Though it can be helpful and great for brainstorming, it's not enough to only talk business over a beer on the porch after a long, hard day. You need to be personally accountable and accountable to each other, so having what we called our weekly "pow wow" helps to keep everyone and everything on track. Decide on how to schedule, manage projects, share files, and keep your internal and external business calendars. You can find many free business tools, like calendars, online. As you grow, you'll figure out whether you need to upgrade to some sort of more sophisticated paid service, such as project management software.

- Determine the importance and urgency for all of your business endeavors. Leadership and time management guru Stephen Covey developed a time management matrix that easily categorizes how you spend your time. You don't want to always be putting out fires or getting behind on important work due to a flooded inbox or spending too much time on social media. Sure, these tasks need to be dealt with, but you want to spend your time consciously. For example, if one of your SMART goals is to inoculate

1,000 shiitake logs by the end of the year, you need to spend time on this regularly.

FIGURE 8.5 Covey's Time Management Matrix with mushroom business examples. Adapted from *First Things First* (Covey, Merrill & Merrill, 1994)

	Urgent	Not Urgent
Important	**I.** The Mushroom Room has become infected; needs to be shut down & cleaned. Your farmers markets aren't gathering the projected profits and cash flow is soon to become an issue. A big show is coming up in 3 weeks, so 200 oyster grow kits need to be prepared. The ad for your annual harvest fest needs to be updated and sent out widely.	**II.** To reach your goal of 5,000 shiitake logs by Dec, you must innoculate 50 per week, on average. To prevent Shroom Room contamination, you must seal the drain and entryway and clean all surfaces. You should meet with that beekeeper about a partnership. You're not sure what operations deadlines are coming up in the next 6 months, so you need to figure it out.
Not Important	**III.** The inbox is getting full and correspondence needs to be dealt with soon. We have a few phone calls to return. We're about a week behind in accounting entries. We have an opportunity to obtain a free logo design from a local college student if we enlist them by Thursday. We haven't posted on Facebook for a few days and need a fresh, creative post to engage our audience.	**IV.** Our house is a wreck and needs to be cleaned. Our social media could be managed better, so we should look for solutions. That salesman called again about a subscription to project management software. Uncle "John" called again to get his free $30 bag of oyster mushrooms. Our display isn't perfect and could use some upgrades. We don't like our bag labels, but still have some to use up.

- Have a weekly schedule of tasks and categorize each into the appropriate time management quadrant, ensuring that you are spending most of your time in Quadrant II and then I, on planning and other meaningful, long-term tasks, such as shiitake log inoculation. There are always plenty of small details, distractions, and interruptions, like responding to customer emails, preparing for markets or an event, and posting on Facebook. Each distraction threatens to take too much of your time. See the figure here for an example of a weekly team schedule, with Covey's time management in mind.

It is also important to develop monthly, quarterly, and annual calendars for "10,000-foot-view" tasks that need to get done, such as tax filings, state annual reports, and various business planning and evaluation

TABLE 8.2. Weekly schedule example.

	Sunday	Monday	Tuesday	Wednesday	Thursday	Friday	Saturday
Dave	**QII:** Business notes; Porch pow wow over some beers	**QII:** Weekly Pow Wow; Shiitake log inoculation	**QII:** Shiitake log inoculation	**QII:** Oyster grow kit prep	**QII:** Oyster grow kit prep; **QI:** Order spawn & supplies	**QII:** Finish oyster grow kits; **QI:** Farmers Market Prep: Harvesting	**QI:** Farmers Market #1
Kristin	**QII:** Business notes & porch pow wow over some beers; **QIII:** Accounting entries	**QII:** Weekly Pow Wow; Meeting with beekeeper; **QIII:** Correspondence	**QII:** Summarize pow wow, send tasks to team; **QIII:** Finish financials	**QII:** Develop operations and planning calendars, share with team	**QII:** Oyster grow kit prep; **QIII:** Update website before big upcoming events	**QI:** Farmers Market Prep: Getting change; Social media postings; Printing	**QI:** Farmers Market #2
Intern 1	Off	**QII:** Weekly Pow Wow; Shiitake log inoculation	**QII:** Shiitake log inoculation	Off	**QII:** Oyster grow kit prep	**QII:** Finish oyster grow kits; **QI:** Farmers Market Prep: Harvesting	**QI:** Farmers Market #1
Intern 2	Off	**QII:** Weekly Pow Wow; Shiitake log inoculation	Off	**QII:** Oyster grow kit prep	**QIII:** Finish headline article; Send monthly e-newsletter to consumers	**QII:** Finish oyster grow kits; **QI:** Farmers Market Prep: Harvesting	**QI:** Farmers Market #2
Kid 1		**QII:** Shiitake log inoculation					**QI:** Farmers Market #1
Kid 2		**QII:** Shiitake log inoculation					**QI:** Farmers Market #2
Kid 3		**QII:** Shiitake log inoculation					**QI:** Farmers Market #2

endeavors. Your email's or computer's calendar functions could serve this purpose. You could develop a business operations or planning calendar through one of these free resources.

Finances

Keep your personal and business finances separate, with respect to both accounting and bank accounts. This is an important factor in limiting your personal liability and maximizing the outcome at tax time. Choose the right accounting software for your business. For very small startups, free cloud-based accounting may suffice initially, therefore maximizing your profit margin. Look at online reviews and then select the software or cloud system that is right for you. Reviews usually include information on pricing, ease of use, and functionality. For evaluation purposes, it might be helpful to know which of your products is most popular and profitable. So, take these factors into account when selecting accounting software.

Budgeting and Balancing

It's important to both project and monitor your finances. This can be accomplished through accounting software that allows you to budget and balance and run various indicator reports, such as monthly cash flow and balance sheet. Though it might require an involved setup process, coding your expenses to their resulting products and services will help you to determine which of your offerings is most profitable and which have too slim of a profit margin. For example, if you sell fresh shiitakes, then identify your shiitake growing materials expenses, such as logs, spawn, tools.

Marketing

See Chapter 9.

Financing

Finding the right funding mix can be tricky with a small startup farming business. Always keep risk in mind and look for ways to maximize your

FIGURE 8.6. Three-year budget example.

Revenue		Year 1	Year 2	Year 3
Sales—Non-taxable items	Fresh mushrooms, fresh vegetables, dried mushroom products			
Sales—Taxable items	Mushroom grow kits, shiitake logs, plants, seeds, cookbooks, artwork, publications, rain barrrels			
Sales Tax				
Services	Workshop fees, consulting, presentation fees			
Memberships	CSA annual membership fees			
Grants				
Loans				
Interest or Portfolio				
Owner Investments				
Crowdfunding				
Community Investments				
Reimbursements or Refunds				
Billable Expense				
Shipping				
Other				
Total Revenue				
Expenses				
Personnel	Payroll, payroll taxes, workers' compensation			
Equipment	Mushrooms, market garden, plant nursery, artwork, rain barrels, office, events			
Supplies	Mushrooms, market garden, plant nursery, artwork, rain barrels, office, events			
Publishing				
Printing				
Fees	State registrations & licenses, conferences/training, trade show exhibiting, farmers markets, accounting software, merchant account, credit card transactions, website, advertising, memberships, banking			
Travel	Hotels, food, mileage, tolls/other			
Consulting or Professional Fees	Tax preparation			
Certified Kitchen or Equipment Rental				
Sales Tax				
Postage/Shipping				
Loan Payments				
Liability Insurance				
Interest				
Household Office Space (Portion of mortgage)				
Utilities				
Reimbursements or Refunds				
Repair or Maintenance				
Donations or Sponsorships				
Discounts or Scholarships Given				
Other				
Total Expenses				
Net Gain or Loss (Revenue − Expenses)				

profits and minimize your costs. Be careful not to get too deep into debt or invest too much of your personal money, especially at the beginning, when you are unsure how much success you'll have.

Paying as You Grow

The best way to ensure that you don't get in over your head is to pay as you grow. Invest what you need to, from whatever source you determine is appropriate, but only what you can afford. You'll likely be getting income from other sources for quite some time, so get into the habit of regularly setting aside profits as they are made for investment into the business.

Crowdfunding

This is similar to fundraising for nonprofits. Specifically it's raising funds for your venture from a large number of people, usually online. According to various estimates found online, crowdfunding revenue volume rose from around $1.5 billion in 2011 to over $5 billion in 2013. And, just from our own observations, we see that it is becoming bigger every year. If executed well, crowdfunding can raise much-needed funds for your business. A variety of platforms for crowdfunding are available; they provide the tools and the third party moderation of the process. With this, too, you'll want to review the various options available and compare them using online reviews. We did not do crowdfunding, but probably should have given it a whirl. A long-time friend of Kristin's who's a fellow farmer tried it, and here's the advice he graciously shared with us:

- It's more time consuming than you probably think.
- You cannot have an "if you build it, they will come" approach. Crowdfunding doesn't work that way.
- Get feedback from friends/family prior to launch.
- Line up donors ahead of time. Get solid commitments from early donors to help build momentum; include face-to-face and social media approaches.
- Have ideas to engage potential donors on a variety of social media platforms (contests, polls, giveaways, etc.).

- Don't assume a specific individual will donate (or not). You may be disappointed (or pleasantly surprised).
- Before choosing a platform, consider its specific perk system and credit card/other fees. Figure this in from the beginning and take into account the benefits/disadvantages of each system.
- Match your perks system with your message. If your business is selling mushrooms, then donate mushroom products to winners of contests, etc.
- Iron out kinks up front, but don't be afraid to let campaign evolve based on feedback/data.
- Much of your campaign time will be used for reaching out/finding new donors. This can be exhausting.
- Consider timing (campaign was launched during holidays, winter…people not thinking about CSA/gardening…however, this can be a great time to get donations, with Giving Tuesday, 1st Tuesday in December, year-end campaigns, etc.)
- Study similar campaigns before you launch; learn from their successes/failures.
- Have a focused message. Don't have a muddled message or try to hit too many "causes" (sustainability, new farmer movement, education aspects, CSA program, etc.).
- Make sure your project can be successful without getting all of the funding you had hoped for.
- Web hits don't equate to contributions. Lots of web traffic/buzz don't necessarily translate into lots of donors.
- Also look into more traditional funding opportunities in local community. There is still a lot of merit/value in doing things the old-fashioned way and hitting the streets, going door-to-door, meeting face-to-face, etc.

Investors, Joint Ventures, Cooperatives, and Partnerships

It's hard to ignore that the wave of the future is collaboration in its many forms. From local food movements to sophisticated joint ventures, collaboration is intended to be mutually beneficial to all parties involved. And, collaboration is not just for nonprofits any longer. Businesses are getting into the game of working together, and doing so can be as mutually beneficial as a tree and its mycorrhizal fungi. Let's look at a couple of simple scenarios:

- To augment our farming operation, we took on consigned honey from a local beekeeper, which was an added income source, but also paired well with our reishi tea recipe, as honey takes the "bite" out of reishi's bitter taste.
- You could add mushroom shares to a local CSA (community supported agriculture) operation, as most farmers do not grow and offer mushrooms as part of their members' weekly shares.
- Participate and volunteer your time with the local food movement in exchange for increased, multi-channel marketing derived from the venture.
- Work with your farmers market compadres to co-market and sell produce used in the same recipes, strengthening each's weekly sales.

The possibilities with various types of formal and informal partnerships are limitless. Regardless of the type of agreement, make sure that expectations are clear from the outset, especially when money is involved or changing hands and when barters are made, such as a two-way exchange of goods or services. It's best to have a written contractual agreement in place and signed by both parties when exchange of goods, services, or cash is involved.

Personal Investment

Here, it's important to minimize your risk by imposing a limit on the amount of your own money you're willing to spend to get your business started. Agree on (and stick to) a percentage of your monthly income and savings so that you don't get in over your head and/or lose equity you've built up over the years.

Government Programs

Low-interest small business loans, cost sharing, and grants can be great ways to finance your business or specific aspects of it. An example of an agricultural cost-sharing program is the USDA's high tunnel program, which funds a portion of installation projects for farmers. Another example is SARE, Sustainable Agriculture Research and Education Program, also supported by USDA. SARE offers various types of farmer grants, which vary from region to region, including support to sustainable agriculture research, and education related to integrated farm systems, organic production, cover crops, renewable energy, and more. You can search for such opportunities at the state or provincial and federal levels, primarily within agriculture and small business agencies. See Resources for more information on government funding opportunities and databases.

Bank Loans and Lines of Credit

Mushroom farming is a risky and difficult business. Try to stay away from loans and credit as much as possible, unless you want *money mayhem*.

Optimizing your Business

Because of the unique challenges of mushroom farming, purveyors need to develop tactics for optimizing success and minimizing failures. You'll need to run a tight ship in order to succeed. Doing so comes in the forms of saving money, making more money, and getting creative.

Saving Money

As a startup mushroom grower, you will need to save money wherever you can. Here are some ideas and things we've done. You will undoubtedly find more.

- Obtain "scrap" hardwood from logging project sites at a minimum cost (or, better yet, free). With shiitake mushroom production, this could save you.
- Grow naturally by inoculating your forest, saving on cultivation supplies.

- Selectively log your property, providing startup money and cultivation medium, both making you and saving you money.
- Arrange to get free or discounted sawdust at your local mill.
- *Only if you are an expert wild collector,* peruse the forest for free mushrooms to sell (see Chapter 4 for more information on wild collection).
- Look for ways to recycle and reuse products you already have in your possession (example: use cardboard as growing medium).
- Retrofit old outbuildings for indoor cultivation (be sure to properly "seal" grow space to prevent or minimize pest infestations and protect the structure's wood framing).
- Minimize labor time by having a schedule for "inoculation days" and get help for those days to build assembly line efficiency.
- Utilize college interns and agricultural apprentices to help you with labor, but be prepared to teach them and provide a rich educational experience in return.
- Select species that are more affordable to grow.
- Make one, take one. Anytime you provide a demonstration or workshop, require your participants to "make one before they take one." If you are showing them how to make oyster grow kits, have them make one for you and one for them to take home as part of the cost of attending the workshop. Alternatively, provide participants with discounts or workshop scholarships if they provide a pre-designated number of hours working for the farm.

Evaluating and Evolving Your Business

Periodically, you will need to check your progress toward your goals. This is where making specific, measurable goals comes in, as well as tracking key indicators for your business, such as shiitake profit margin. When making your goals and setting up your accounting practices, keep in mind that you will need to look back later and evaluate your progress so that you can adapt and adjust as necessary. Along the way, you will encounter both challenges and opportunities, so be sure to pay attention to how these impact your business. Some challenges come from within, so you will need to make operational adjustments accordingly.

Some obstacles are external. Therefore, you will need to do whatever is within your power to navigate through, around, or over these challenges. Pay attention to your own operation, as well as outside forces, such as your changing markets. Transition your business to successfully deal with change, create the change, and thrive. According to Covey's time management matrix and your monthly, quarterly, and annual schedules, take a look back and ahead on a regular basis. This is a vital Quadrant II—important and not urgent—task.

Supplemental Income
or Value Addition to Existing Business

Can you answer yes to any of these questions?

- Do you have a wealth of hardwood resources, sawdust, straw, or scrap logs available to you at your home or business location?
- Are you a farmer looking to diversify your product mix?
- Are you an employed person or business owner with the freedom (time and money) to experiment with the addition of mushrooms as supplemental income?
- Are you successfully growing mushrooms for yourself and would like to know how to sell excess locally?
- Are you an active person who is searching for a retirement business?

If you answered yes to any of these questions, you may want to consider adding mushroom cultivation to your suite of products OR work with mushroom growers to sell more of your existing product. Regardless of your particular situation, you should consider risk. How much time and money can you safely invest in order to try mushroom sales? You must figure out your appropriate formula for "going all in" and minimizing risk.

Though your growing methodology is the same as someone who is seeking to focus solely on mushroom cultivation as a business, your other considerations are significantly different. You are not hoping to focus solely or primarily, or even in large portion, on mushrooms as a business. Therefore, your risk is lower.

Here are some examples of how to incorporate mushrooms into your existing business plan:

- If you have a byproduct from your business that is useful to mushroom growers but is normally not a money maker for you, or you'd like to sell more of it, connect with cultivators in your area to establish a relationship. Some byproduct examples include sawdust and scrap wood from a logging site.
- If you own a family farm, incorporate mushrooms slowly to find out what works for you in terms of time and money needed. If you have straw, you could start with wine cap stropharia cultivation in your market garden. If you have plentiful hardwoods that are appropriate for growing shiitake, you could start with shiitake logs.
- If you already grow mushrooms for yourself and would like to sell the excess to more than just friends and family, start with your local farmers market or target restaurants or health food stores. Note: You will need to consider officially starting a business to protect yourself, so make sure it's worth the trouble. And, with restaurants and retail shops, you will likely be required to have business and liability insurance.

Regardless of your unique situation, you should take it slow, proceeding thoughtfully. You might want to consider perfecting the cultivation and sale of one or two varieties of mushrooms and derived products at first, and then proceed to adding a diversity of products over time. Once you know what you would like to start with, you can pick and choose applicable information from this book and the Resources section, much like a buffet.

Marketing Mushrooms

"What do you have here?…Ohhh…eewwe, mushrooms, yuck!" gripes our first potential customer of the day, with an overly frightened expression, as if the mushrooms were about to jump off the table and attack. David and Kristin look at one another with the understanding that only seasoned farmers market mushroom vendors share. It's going to be a long day if this keeps up. Second visitor…same reaction, but with a bit more tact, thankfully. Though the facial expression was about the same. We sure do hope that our regulars, Richard, Larry, and our new friend, the woman with the long, flowing hippie dress, aren't on vacation this week or we'll be eating all of our profits—literally. Kristin daydreams, "Mmmm…shiitakes in a nice hardy stir fry sounds awesome." David's thoughts parallel, "Maybe the wine caps won't sell and we can sauté away!" We snap back to reality, pleading with the inner mushroom lover in each of us, "Get rid of those thoughts…We need to make *some* money at this!"

Mushrooms should market themselves, right? They're tasty and can contribute to a long, healthy life. They should be flying off the shelves, or off the farmers market folding table, at any rate. They can be used in a multitude of ways and have a wide variety of flavors and benefits. Win-win. The mushroom farmer is happy and so is the customer. With your die-hard mushroom lover—the weekly returning customer—the ease of selling mushrooms is evident. But, marketing mushrooms is not always that easy.

In the local mushroom selling business, you can't just count on success by farmers market. Some venues' customers are more into mushrooms than others'. Mushroom growers nearly everywhere need more than the farmers market. If you want to sell mushrooms locally, you will need to find the mushroom lovers and make it easy for them to find you. Then, you will need to develop and care for relationships with those fellow *shroomees* to protect the resilience of your business and to enrich your life.

When it comes to mushrooms, people either love them or hate them. There is no in-between. You must connect with the local mushroom lovers to have a chance at succeeding with your new mushroom farming endeavor. A number of challenges can wreck your mushroom growing business, whether it's a small or large operation, whether it's in a good market or a more challenging one. Some of these challenges are unique to mushroom farming, while others are struggles shared by other types of farming operations and all kinds of small businesses.

In this chapter, you'll learn some tried-and-true small farm and mushroom-specific marketing considerations and how to develop your own marketing strategy, one that works for you in your unique market. When it comes to marketing, you can't do it all (even though a buffet of ideas and options are offered here). You're likely limited in time and money, perhaps severely. So, knowing what will work for you, within your market, at the various stages in your growth will help your mushroom business immensely, whether only growing mushrooms or adding them to your existing mix of farm products.

Though the buffet of marketing options offered here is probably too large (like most buffets), a thoughtful strategy will empower you to focus your limited time and money toward effective marketing endeavors. The ultimate goal of your marketing efforts is to create a strong, long-term customer base, a mycelial-like network of mushroom lovers who support your business. Just as the fruit of a mushroom needs its extensive mycelial network before it fruits, your business needs a strong, extensive network to be fruitful. Therefore, the options here will flow from a single trunk of mycelium, out to the many smaller branches. Your trunk is the foundation upon which you build your marketing strategy, which continually branches out into new and exciting territories. As time progresses, you'll have a fruitful and resilient business.

In this chapter, we cover these topics and offer resources for:

1 **Challenges and opportunities:** For small farm businesses and selling mushrooms, in particular.

2 **Developing your story:** Branding, testimonials, and other content you need, whether marketing online, through paid advertising, in-person, or all of the above.

3 **Traditional advertising:** Print materials, newspaper ads, and other tried-and-true methods that remain relevant today, but must be approached differently in the modern world.

4 **Target marketing and understanding your unique market:** Especially if your local market presents unique challenges, this section presents ideas for reaching those mushroom lovers, beyond the farmers market.

5 **Crafting your online presence:** General guidelines and specific ideas for navigating websites, social media, and the endless options for getting noticed by the right people online.

6 **Collaboration:** Ideas for how you can work with farm businesses and other partners to make everyone's marketing efforts stronger.

7 **Face time:** Amping up your effectiveness through face-to-face interaction with people through participation in on-farm and community events.

8 **Developing your unique marketing strategy:** A vital part of any business plan, your strategy will guide you into the future, and can be adjusted at any time

9 **Evaluating your efforts:** Knowing what is and what isn't working is important so that you can make necessary adjustments.

10 **Looking forward and anticipating change:** People change. Markets change. Businesses must also change. Be ready by keeping an eye on present and future trends.

By no means is this chapter a comprehensive course in small farm marketing. Rather, our intention is to expose you to marketing tactics that have worked for us, lessons we've learned, a buffet of marketing options, mushroom-specific marketing considerations, and the resources you need to find the marketing avenues and strategies that work for you. Branch out beyond this chapter by exploring the related resources provided, which will likely lead you to even more tools you can use.

Challenges and Opportunities
Mushroom-specific

The Shelf-life Challenge

First, let's look at challenges unique to small-scale mushroom farming operations. Mushroom fruiting is somewhat unpredictable, and distribution needs to occur immediately due to mushrooms' short shelf life. This reality presents a unique marketing challenge. If you can't create a viable plan to get your mushrooms sold right away, you should stick with hobby growing until something changes (which wouldn't be so bad, right? Mmmm...shiitake stir fry!).

Found by us to be the easiest method, a weekly post to Facebook will let your farmers market customers know a day or two in advance what will be available at the upcoming market, or you can post whenever something has recently fruited and is super fresh. A post could invite customers to a pick-your-own party during specific open hours when a big flush of mushrooms pops up. You can also make posts on your website and use email blasts to alert customers to available mushrooms, if you have the time to do so. If you want to let people know about products in multiple forums but are limited in time, consider posting to one forum that automatically feeds to the others. For example, we sometimes posted product availability via our e-newsletter, which was set up to post to our Facebook page, which streamed live on our website through a social media plug-in. An established working relationship with a local restaurant can provide you with a great place to unload fresh mushrooms in a time crunch, especially when the restaurant is a farm-to-table establishment and/or the chef is flexible enough to make immediate magic with an unexpected supply of mushrooms.

Love, Not Hate

You must find the mushroom lovers (and be prepared to politely ignore the mushroom haters that stop by your farmers market table and say things like, "Eew, slimy, gross mushrooms!"). Especially in some geographic locations, you're not going to find all of the lovers at the farmers market. So, where can you find them? It depends on where you are, but here are some examples to get your creative juices flowing. Look for the lovers in these places:

- The local mushroom club
- Ethnic clubs
- Cooking classes or clubs
- At local celebrations or festivals, such as heritage days

See more in the target marketing part of this chapter.

Lack of Trust

Trust is important with any small farming operation. But when selling mushrooms, trust is critical. People can be wary of mushrooms and their safety, even cultivated varieties. So, you can build trust by being knowledgeable and sharing your knowledge, developing friendships, and being transparent with your operations. Educating customers on the safe consumption of mushrooms will serve your business well. For instance, you can advise customers to start with small quantities of a new kind of mushroom to ensure proper digestion. You can show that you are sensitive to the fact that some people have a sensitivity or even an allergy to mushrooms. You can share your safety procedures and your mushroom spawn sources. Most importantly, you can distinguish between your cultivated and wild-collected varieties, if you collect and sell mushrooms from the wild. These actions will build trust in you and your mushrooms.

Small Farm Business

Limited Time and Money

You will likely have very limited time and money for marketing. You may even admit that you don't have a clue how or where to market your products—other than the roadside stand or farmers market. This chapter is designed to take some of the guesswork out of marketing your mushrooms.

Friends, Not Customers

It's not just about marketing. It's about building good, lasting customer relationships, which can also be your best marketing tool. Since it's a lot harder to gain a new customer, keep the ones you've got. How? Engage them. Have conversations with them. Ask them how they prepared the shiitakes you sold them on Saturday. Ask for input and incorporate their suggestions. Consider incorporating open forums and other types of two-way conversation avenues so that you can stay in touch with them. Not only will they come back for more maitake, but they will tell others how awesome you and your mushrooms are. You may even learn some valuable lessons from them that can be incorporated into your business toolbox. More importantly, good relationships are an essential life asset that enrich you deeply. We view our mushroom business as an extension of who we are, as a joyful social and environmental venture, in addition to serving as part of our livelihood. If you, too, view your business this way, it will be fruitful, like shiitake on an oak log.

Developing Your Story
Branding and Graphic Design

If you're starting your farming business from scratch, or if you just haven't had the time to dedicate to ensuring that your marketing strategies are effective, you should start with developing your brand. What logo, images, tagline, or overall look and feel will distinguish your farm so that your products are easily recognizable upon first glance?

Overwhelmed? You're not alone. You're probably thinking, "How am I to develop my brand when I know nothing about branding?" Here are a basic checklist and some ideas to make it a lot easier.

Checklist

- Develop your vision, mission, and values statement.
- Develop a simple, easy-to-recognize logo.
- Decide on a 3–5 word catch phrase or tagline that embodies your business.
- Select 3–5 high-quality images to use repeatedly in your marketing.
- Based on the colors of your logo, select a complementary color scheme for your website and other marketing materials.
- If desired, select 2–3 fonts to be used consistently within your marketing materials, using each, respectively, for headings, sub-headings, and body text.
- Develop *infographics* on the benefits of mushrooms or other data to support the need for your products.

FIGURE 9.1. Berglorbeer Farma logo.

Logos and Photos—A Few Ideas

- Engage a local college marketing or graphic design student in branding development, as part of a semester-long internship or class project.
- Hire a graphic designer at very little cost to develop your logo at a "nearly free" online site.
- Go it alone if you or someone in your inner circle has the creativity and time.
- Use a DIY graphic design site to develop your branding style or use in combination with your own software. Even basic presentation or publishing software can generate quality graphics if you know how to manipulate them (but be careful to pay attention to copy-right agreements for commercial use, especially with free or cheap online services).
- Take your own pictures to enhance your flexibility and credibility.
- If you do not have your own pictures yet, search for free or low-cost stock images or ask permission to use other growers' photos, with credit provided to them.

FIGURE 9.2. Social media share contest graphic.

Featured here is a social media marketing image we created in less than five minutes using an online design site.

Testimonials and Other Desirable Warm, Fuzzy Things

Fuzziness on your oyster mushroom growing medium may indicate a serious contamination problem. But warm and fuzzy marketing should have a place in your endeavor. People are inspired by both facts and stories. People can be prompted to buy your mushrooms because of their proven health benefits. But they can also be encouraged to buy by your unique farm story and stories from your customers who have benefitted from your services and products. As a business, you need "standing" or previously developed content that you can use over and over again in your various marketing endeavors. Standing content comes in the form of stories, logos, testimonials, and more.

First, develop your unique story. Who are you? What do you care about? Why farming? Why organic methods? Why mushrooms? Here are the basics of developing a good story about you.

Good storytelling:

- Has a great start…something relevant and amazing that happened, such as an event that brought you to mushroom growing OR an inspirational quote that motivates you and is related to your unique story.
- Has emotion; is compelling.
- Conveys your primary messages—your values, your superior product, your goal, how you care about your customer.
- Is personal; speaks informally to the customer.
- Is simple, short and sweet. A paragraph might be enough, a couple is certainly enough.
- Is authentic and genuine (non-fiction)—does not exaggerate the truth or make up compelling stories.
- Has a FEW compelling facts or statistics, such as mushroom health benefits.
- Answers: "Why mushrooms? Why this cultivator?"
- Contains pictures and other graphics.
- Includes a testimonial from someone similar to your target audience or someone they can relate to.
- Should have humor somewhere in there.

Your Standing Content Checklist

Below is a checklist of standing content you will want to develop and have on hand when marketing your goods and services.

Branding:
- Logo
- Color scheme
- Images
- Fonts
- Vision, mission, values, and tagline

Narrative:
- Your story
- Inspirational quotes
- Testimonials and reviews
- Ambassadors
- Your presentation
- Infographics
- Your recipes

Testimonial Example from Our Farming Business:

"I love working with folks at Berglorbeer Farma. They offer a beautiful and consistent product. Dave is happy to work with me on exotic mushrooms for special dinners as well as providing the best oyster mushrooms I have ever tasted for my a la carte menu.

"I always look forward to the new and delightful fungus that Dave is cultivating or foraging!"

—Jane Doe, Executive Chef, Awesome Local Restaurant

Recipes

A recipe book is a great way to market your mushrooms. Established publishers and online DIY recipe book creators (which provide the formatting, layout, and specifications through their online program, while you enter all content, resulting in a printed recipe book, for a fee) are both options, or you could self-publish your book. Regardless of how you publish your recipe book, time is what you will need most for this endeavor. It will take longer than you think it will, so budget your time accordingly.

Education

Educating the public about the benefits of mushrooms and their value as tasty culinary ingredients takes time and much effort. But if you're in it for the long haul, providing education is worth exploring. Here are a variety of ways you can provide education:

- A twist on the concept of a recipe book: develop a calendar featuring a different culinary mushroom each month (with information and a recipe for each) to sell.
- Develop a presentation you can give repeatedly to different groups.
- Write and distribute an article to various media outlets.
- Consider pitching a recurring newspaper column on healthy, sustainable, local foods to your local paper. Because this is a big commitment of time, be sure that you have the time to follow through on writing your column. You may want to start with a monthly column, rather than a weekly one.
- Write a blog.
- Post recipes to your Facebook page.

Traditional Advertising

Traditional advertising, for the purpose of this section, includes print advertisements, other paid advertising, and free advertising options. Consider checking into these options for advertising your products:

- Buy/sell Internet sites like Craigslist or Facebook-based classifieds. Though some do not allow the sale of food, you can market and sell your recipe book, mushroom logs, or grow kits on any of these sites for little or no cost. Each has its own rules and costs for business advertising.
- Buy/sell printed circulars.
- Newspaper ads. Get together with other local farms on a shared ad, perhaps leading up to harvest or holidays.
- Get yourself mentioned in newspaper articles, such as those that feature local people doing great things or regular columns with public interest pieces (think: what is unique about what I'm doing, and to whom can I pitch the article idea?).

FIGURE 9.3. Advertisement example.

Printed Marketing Materials

Costly? Usually. Effective? Usually. Printed marketing materials can play a key role in getting new customers and increasing your returning customer pool. At events, especially more involved ones like farm festivals, effective signage can attract people to your exhibit, while printed materials can help attendees find you after an event. If carefully thought out, printed materials do not need to break your marketing bank. For instance, consider business cards that double as recipe cards. A customer is much more likely to keep your business card if it has a tasty recipe on the flip side. A brochure doesn't have to be a full page tri-fold. It could be more like a rack card, which is a third of a sheet. A banner is a one-time cost worth pursuing if you plan to travel to multiple events. Once produced, no further spending is needed. Plenty of online printers offer great deals on marketing materials, such as business cards and brochures.

Here is a list of printed materials to consider producing:

- Brochures
- Business cards
- Recipe cards
- Banner
- Signage
- Swag (various goods, like t-shirts, mugs, mushroom storage bags, etc. with your logo, tagline, or other branding images)
- A QR code (quick response code) sign, with the QR code linking on-site smart phone customers to your e-newsletter subscription link, for instance (a growing number of customers have QR code scanner apps on their smart phones, and if they have this feature, they can snap a picture of your QR code, which can take them anywhere you want them to go, such as your website, your e-news sign-up, etc.)

FIGURE 9.4. Brochure example.

FIGURE 9.5. Dave's business cards at the market.

Target Marketing and Understanding
Your Unique Market

You should first ask, "Who are my potential customers?" Once you answer that question, you need to figure out how to find them. If attending local farmers markets doesn't help you expand your mushroom business as much as you'd like, consider diversification into non-perishable mushroom products that can be shipped anywhere, such as DIY mushroom kits or dehydrated products. For most small farms, though, the local market is key to success. To find mushroom lovers locally, beyond the farmers market, search in these places:

- Local mushroom clubs
- Ethnic clubs
- Other community venues
- Wellness or health classes
- Naturopathic/homeopathic/holistic doctors and clinics
- Health food markets and integrative health/medicine centers
- Cooking, vocational-tech, agriculture, nutrition, and community education classes
- Community garden initiatives and garden clubs

Leveraging Your Organic
or Sustainable Methods as Marketing Tools

People are attracted more and more to products that will help them achieve the goals of being healthier and living longer. Wholesome, chemical-free food is near or at the top of the list of products that help people achieve health. Therefore, not only will you use mushrooms' nutritional information to market your goods, but you also want to tell the story of how you use sustainable methods, how you grow the mushrooms organically, or otherwise how you are making a difference in the lives of people and the environment through your operation. Increasingly, people will support the socially responsible business over one less concerned with such matters. Include this information in your story.

Two-Way Communications with Customers

Don't just speak at your customers. Invite them to join the conversation about mushrooms, about your business and their experiences with you. Start dialogues on a diversity of related topics. Use your website, blog, and social media to converse with them. Become friends with your customers and listen to what they have to say. Engaging them reminds them of your presence and gives you an opportunity to learn from them and make suggested adjustments. Have fun and be creative in your conversations. One of many ways to do this is to ask your Facebook followers what their favorite mushroom is and why. If you are having trouble engaging your crowd, then provide incentives for participation from time to time, such as a chance to win an oyster grow kit.

Word of Mouth

Be sure that your customers are not only happy with your products, but also with you. Their dealings with you can make or break you. Why? Because word of mouth is still a very powerful marketing tool, maybe even more powerful now than ever before. Not only do people still get together and talk face-to-face, but now, they also communicate more frequently with a much wider audience through social media. If they love you and your mushrooms, they will tell their friends, who likely have a similar interest in good, gourmet, healthy foods. Give returning customers special deals at the right times, such as when they are hosting their friends at a holiday party featuring your mushrooms. Consider developing a refer-a-friend discount program, where the referrer receives a discount coupon for adding to your customer base.

Crafting Your Online Presence
Websites and Email Marketing

Website Developers and Builders

These days, there are so many ways you can build a website...Weebly, Wix, Wordpress, Small Farm Central, a professional developer—just to name a few. Ugh! How can you possibly choose the one that works for you?

First, determine how much time, money, and technical expertise you have to offer your website development project. (See Figure 9.6 showing website platform decision matrix.) Then, determine which sections, pages and features you'd like to incorporate on your site, such as:

- A home page slideshow of your best pictures
- A plugin of your Facebook feed
- Videos
- Your blog
- An online store
- A calendar of events
- Search engine optimization
- Ability to manage content internally
- Google advertising credits, such as Google AdWords included

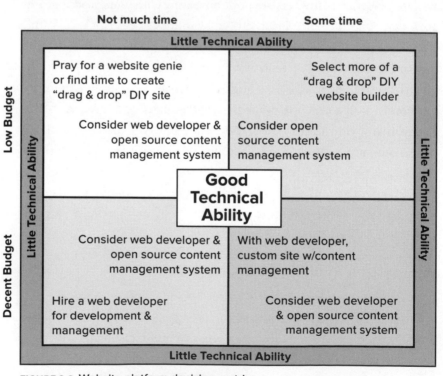

FIGURE 9.6. Website platform decision matrix.

Do your research using online website builder comparison tools or conversations with website developers. Determine, based on all factors, which website development process and platform works best for you.

Whatever methods and platforms you choose, make sure that your site can be easily found by your target audiences. For instance, if you are selling DIY mushroom kits to anyone in the lower 48 of the United States, you must have good search engine optimization so that when people search "mushroom kits," your site shows up near the top of the list. If you exhibit at Mother Earth News Fair or other such events, make sure that the event site links to your site so that attendees find you outside of the event. The more you're "linked up," and easily found through a web search, the better.

Email Marketing

A great way to connect with your customers is through production of an e-newsletter, provided that you have the time to put it together and distribute it on a regular basis, at least monthly. Unlike websites, e-newsletters do not have to cost any money at all, as long as your

FIGURE 9.7. Berglorbeer Farma website home page.

subscriber list is under about 2,000 people. After this threshold, you will have to pay a fee. If you plan to grow your list beyond 2,000 (and you should!) compare the platforms that offer free plans to learn how many subscribers will require a paid plan and what those costs will be as you grow.

Here are some guidelines for effective email marketing:

- Direct people from your e-newsletter to more information on your website.
- Choose a platform that is easy to use, aesthetically pleasing, and can handle large group emails—some of these are even free up to a certain number of subscribers.
- Consider selecting a platform that offers a downloadable QR code you can print so that people can sign up at events simply by scanning the code.
- Honor your customers' privacy.
- Never add people to your subscriber list without their permission.
- People get a lot of email, so be creative with your subject line to catch their attention, but also try not to bombard them.
- Keep your e-newsletter content interesting but concise. Use pictures and creative language. Consider recurring topic categories, such as "Featured Fungi."
- Use incentives, such as, "Sign up and receive our list of the top 5 culinary mushrooms."

Here's an example of what we would tell potential subscribers:

"Sign up for *Cap Chronicles*, our e-newsletter. Receive updates on our products, services, and events, as well as useful information on sustainable living. We publish *Cap Chronicles* about once a month and periodically send special announcements. We do not bombard your inbox, and you can instantly unsubscribe at any time. We never share or sell subscriber contact information."

Connect with Us

Visit our Website
www.berglorbeerfarma.com

Like us on
Facebook

Scan the QR code to
sign up for *Cap
Chronicles* E-News

FIGURE 9.8. "Connect with Us" graphic from display.

Social Media

Start with Myspace…Just kidding. Start with Facebook.

A great number of people in your target market are likely active Facebook users. Mostly, they are in their 30s, 40s, and 50s and love good food. Just take a look at all of the food and recipe posts from your friends who use Facebook. If you are severely limited on time or patience for social media (or knowledge about it), you should focus solely on developing your Facebook presence. Facebook can be a wonderful way to market mushrooms, as you can post, in real time, what is currently available. If you have a website that is mostly static in nature, you can insert a Facebook feed onto your website so that visitors to your site can see what's available whether they use Facebook or not. A weekly post prior to your farmers markets can boost your market sales for the week. Posts, at any time, about available products can help you sell your mushrooms before they go bad. A Facebook post is also an effective way to let your customers know about specials or deals you have running.

Some other common social media tools used for business include Google+, YouTube, Pinterest, Instagram (a foodies' favorite) and Twitter. As with website development platforms, comparison tools and information about social media for business can be found online. Also, social media management tools can help you more efficiently post content to more than one social media platform at once.

Be Creative! Contests, Cartoons, and Other Fun Strategies

As time permits, have fun with your social media tools. It can be as simple as posing a question to your audience, such as, "What is your favorite mushroom?" or "Guess which mushroom on our list has the highest immunity benefits?" You could tell about a product you sell on "Mushroom Monday." A great way to build your Facebook audience is to host a share contest, where you challenge your friends to

FIGURE 9.9. Facebook contest example.

share your page with their friends, which you can monitor. Each friend who shares your page is entered into a contest to win a product or discount from you. Caution: Pay attention to social media marketing and contest rules.

Even if you cannot draw well, you can create a funny cartoon about mushrooms using a free online cartoon creator or simply use one of your pictures in the post. You can post notices about a sale you're running using a free online graphic design tool. With these free tools, it may not always be possible to download a picture file of the creation, so learn how to take screen shots of your creations and save them to your photo folder.

Collaboration

Like the spores of a mushroom traveling with the wind, spread the wealth and collect the benefits through collaboration, including business marketing ventures. There is power in numbers. Ten small farms can buy more advertising space than one. One small farm is unlikely to build the local food movement, but a local food network of 50 farms can effect change successfully. The goal of true collaboration is to lift up all involved for mutual benefit. Just as trees need mycorrhizal fungi and vice versa, collaboration should focus on mutualism. Collaboration is the wave of the future, as small businesses increasingly rely on one another for mutual success. Sure, competition will always exist. But, more success is derived by collaboration.

Cross Promotion and Food Pairing

Through the lens of mutual benefit, small farm partnerships can strengthen all involved. Think in terms of food pairing, to start. Look through the recipes that feature your mushroom products. Most likely, you have one involving mushrooms atop a steak. Mmmm…shiitakes and onions on grass-fed steak! You probably have recipes that focus on vegan or gluten-free dishes. Ask yourself, "Who is producing the products on these recipes that I do not produce myself?" Find those who do and connect with them. Ask them about the possibility of cross promotion. Locally, you could develop recipes that call for "1 lb. of *Moo Farms* grass-fed steaks." As you'll find in this section, a number of avenues of interaction with other producers could already be available locally. That's not to say that you cannot reach out to others on your own. Who knows, it may result in a new CSA or local food network.

CSAs

Community supported agriculture (CSA), according to the USDA, is "a community of individuals who pledge support to a farm operation so that the farmland becomes, either legally or spiritually, the community's farm, with the growers and consumers providing mutual support and sharing the risks and benefits of food production. Typically, members or 'share-holder' of the farm or garden pledge in advance to cover the anticipated costs of the farm operation and farmer's salary. In return, they receive shares in the farm's bounty throughout the growing season, as well as satisfaction gained from reconnecting to the land and participating directly in food production." (USDA, 2014)

The potential for food pairing and cross promotion within multi-farm CSAs is vast because a CSA customer share could include nearly every ingredient needed for multiple meals. Most CSAs do not include mushroom shares, unless the CSA host farm grows mushrooms. A lot of CSAs are single farm products. However, multi-farm CSAs can be formed as well so that the customer receives a wider variety of products within their share. You could find a single farm or multi-farm CSA and approach them about adding your mushrooms to their CSA distributions as they are available. The CSA host farm could pay you a wholesale price for your mushrooms. You could also form a new CSA through a crowd funding platform.

Restaurants and Caterers

Increasingly, locally-owned restaurants are incorporating local, seasonally fresh ingredients into their menus. More and more farm-to-table restaurants are appearing. Establish a connection with a flexible chef who frequently cooks seasonal specials or revolving dishes. Schedule an appointment, and when you go, bring the chef at least a half-pound of mushrooms. Advocate for a menu listing of your farm's name, citing the increasingly aware consumer of local products, and local foods as a marketing tool for locally-owned restaurants.

Local Food Networks

Increasingly, local food lovers, producers, and purveyors are coming together, organizing to strengthen the local food movement in their communities. Though these groups take various forms, the aim is the same: to strengthen family farms, connect people with local farmers, and promote healthy people and local food economies. The groups work together toward these common goals by collaborating on a wide variety of objectives toward their purpose. Their efforts include projects like:

- Collaborative marketing, i.e., marketing cooperatives
- Connecting farmers with markets
- Organizing and managing CSAs, new farmers markets, and other distribution and market mechanisms
- Developing a collective voice on local food issues and policy
- Educating the public on the importance of local food

Some groups are independent, while others are formed as part of a larger effort, such as the Buy Fresh, Buy Local program. Either way, groups often participate in or even develop means by which connections are made between local farmers and consumers, such as websites, social media pages, group advertisements, or listserves. Most have a paid membership structure, which enables the group to meet its mission and provide benefits to members, such as increased exposure or group advertising. Because the existing food system heavily favors industrial agriculture, these groups are not just a nice idea, but rather a necessary community group, if the local food movement is to gain the traction it deserves and people need.

But local food networks have spotty coverage, can be called by various names, and are promoted and aggregated by multiple entities and their initiatives. Therefore, they can be difficult to find and understand. To help dispel confusion and assist you in finding a local group near you, terminology and resources for similar, but not equivalent, groups advancing local food include:

- **Local food network:** a common, general term for groups working collaboratively to advance local food markets.
- **Buy Fresh, Buy Local Chapter** of Food Routes Network, LLC: a membership-based program of Food Routes Network, LLC, initiated by the Pennsylvania Association of Sustainable Agriculture.
- **Regional food hub:** "a business or organization that actively manages the aggregation, distribution, and marketing of source-identified food products primarily from local and regional producers to strengthen their ability to satisfy wholesale, retail, and institutional demand." (USDA Agricultural Marketing Service, 2012)
- **Foodshed:** "The idea of a foodshed is based on a watershed, which is that ecological area connected by water flowing from brooks to tributaries, to rivers. Instead of water, it is the flow of food that defines a foodshed. Foodsheds are technically geographic areas that you could draw on a map, much like the radius of a certain distance around a city. But more accurately, foodsheds are places that connect common food and agricultural interests through commerce." (National Good Food Network, 2015)
- **Marketing cooperative:** a general term to describe a group effort to market collective goods; could be organized in the form of a local food network, CSA or other structure.

Through various efforts by local, state, and federal agencies, a number of websites host information on where to find local food. For producers, registration and inclusion of your farm and products available is free on a number of these sites. But membership fees apply for others. Here is a list of online resources that connect people to sources of local food and to local food networks:

- Buy Fresh, Buy Local
- Local Harvest
- Rural Bounty
- Farmers' Web
- Farm Match
- Food Hub
- Know Your Farmer, Know Your Food
- ATTRA Directory of Local Food Directories

See Resources for more information.

> **Tip:** To find other local farm and food resources, don't
> search "find a farmer" unless you're single and looking to
> enter the life by marriage. The appropriate marketing sites,
> like Local Harvest, appear when searching "find a farm."

Local Business Promotions or Buy Local Campaigns

Now more than ever, buy local campaigns are spreading. They are organized by a variety of business support entities, such as chambers of commerce, local business alliances, downtown partnerships, economic development agencies, and coalitions of local businesses. Because collaboration is increasingly important and there is power in numbers, any local food network effort should strive to connect with buy local campaigns, which usually include all local business. Ensuring that buy local and local food campaigns are coordinated can create a powerful boost to your local food movement, as buy local campaigns often possess more resources than local food networks, such as advertising dollars. Plus, it just makes sense to work with your buy local campaign, as you, the local farm, are part of the local economy, just as the downtown café is. Do some research to find out if a buy local campaign exists or is being formed in your area and work to pair the local food network with it for more power and better results.

Face Time

Unless your business model is to sell mushroom products only online, no website, social media tool, or e-newsletter can replace or be more impactful than face-to-face interaction with people. Continuing relationships are formed most successfully in this way. Eye-to-eye contact, a friendly smile, a good conversation, and a firm handshake go a long way to engender trust in you and your products. Plus, people at food or sustainable living events are your target market. They are there because they are seeking local food or the knowledge they need to grow their own food. We found some of our most loyal customers and friends at events such as farmers markets and farm festivals. Almost 100 percent of our mushroom-growing workshop attendees were found at face-to-face events—usually at our exhibit. You'll find that when you run online campaigns, from simple things like social media share contests to large crowdfunding campaigns, the more community connections that you have from face time, the more successful your online efforts will be.

Agritourism (or Agrotourism)

Parallel to the local food movement, people are increasingly visiting farms, both locally and when traveling. According to the USDA Agricultural Marketing Service, *agritourism* (or agrotourism) "describes the act of visiting a working farm or any agricultural, horticultural or agribusiness operation to enjoy, be educated or be involved in activities." Examples of agritourism are:

- Farm tours for families and school children
- Day camps
- Hands-on chores
- Self-harvesting of produce
- Hay or sleigh rides
- Overnight stays in a bed and breakfast
- Harvest festivals or open houses on your farm
- Workshops and classes

To really connect with your customers, consider developing some onsite programming so that they can see where their food comes from and how it is produced. As with other examples of face time, this engenders trust and friendship with your clientele. But, agritourism goes beyond face time at the market by proving to your customers what a great operation you have, how clean it is, how wonderful your sustainable practices are, how delicious your mushroom recipes are, and even how much of a "fun guy" you are. Be prepared to show off, though. It's all about transparency! Provide good directions and even signage. Greet people as they arrive. Shake their hands. Make sure all of your sustainable practices are in good working order and can be easily shown. Be sure your place is very clean. If you have an indoor growing facility, show your visitors, but be sure that you take precautions to avoid contamination, such as instructing visitors not to touch any surfaces. Serve good mushroom dishes or samples, always including vegan and gluten-free options. Keep the kids in line…you don't want Dennis the Menace mayhem. Have music or invite people to bring instruments.

Take care when inviting people in. We have attended on-farm events that turned us off from those producers because they didn't demonstrate that they were top-notch operations—some were far from it. Kristin once stayed overnight at a farm where she had trouble finding a spot to set up her tent because of the quantity of fresh dog piles all around. Don't be that farm.

At the Farmers Market

One word for the novice about selling at farmers markets: Prepare! Farmers markets remind Dave of his youth when he followed the Grateful Dead. Just like the Grateful Dead shows, the people at a farmers market make up a unique subculture. You will soon learn the personalities of those around you. You have to remember that you are there to make money. The reality is that this method of making money is a lot of work, will probably not generate sufficient income alone, and you will meet every type of personality!

Even if you followed the Dead, like Dave, don't wear your Jerry Garcia tie-dyed t-shirt to the market. Even without it, you'll receive this question, "Do you sell those funky mushrooms?" You'll want to answer (and sometimes might!), "Why yes, I do bring illegal mushrooms to the public farmers market, but I keep them behind the counter awaiting complete strangers to walk up and request them, because I sure do hate my current life and can't wait to meet my cellmate for the next 10 to 20 in the federal pen!" Plus, you'll already be dealing with this too-common and perhaps most asked, preposterous question: "Are these poisonous?" You'll want to answer "Why, yes they are. We don't need repeat customers." Or "Yes, they're poisonous because I also sell caskets as my second income." Yet another question you'll get, "Are they edible?" Again, you'll want to say something smart, like, "Nope, we just sell them as a craft supply." These are all questions we heard over and over again. So, be prepared, stay calm and try not to succumb to the urge to deliver any of the above replies.

Admittedly, we often reply with sarcasm to the funky mushroom question. Sarcasm can be a wonderful tool, but it has to be used carefully. Use your senses about people to determine the correct response. Your notions of human evolution and intelligence will be tested, so try to have a sense of humor—most of the time. Being completely rude is almost never appropriate. But, once in a while you have to break down and let people have it right between the eyes. It's just human nature and sometimes, it's justice. Once, a lady stood at our booth for quite some time, talking to a crowd of our potential customers, saying, "Mushrooms are disgusting, I hate them. They are poisonous. I once ate them and almost died." Dave responded, "I asked my mushrooms, and they don't appreciate un-educated opinionated people. Look lady, I'm trying to feed my family, so why are you standing in my booth ripping my product with your opinion in front of my potential customers?" You have to have thick skin, but you don't have to be defenseless. We estimate that 75 percent of people attending a farmers market are looking for healthy food or new products; 23 percent are there for the social aspect of it; and 2 percent of attendees are just sour about life or are lost and wandering around.

Concentrate on the 75 percent, as they are your core clients. Be prepared to educate and help them with their purchases. Take a few moments and talk with them to get a feel for what they are looking for. A burnt lion's mane probably won't garner repeat customers, but a quarter pound of oysters with a recipe card usually does. Once people have a good experience with you and your products, they will be back next week, maybe wanting to try shiitake this time. We can't tell you how many times our variety of products helped us to create and keep customers as friends.

A few words about other vendors: Whether it's the racist flower peddler you dread having to listen to each Saturday morning or the family farmers down the road that you enjoy talking to, once again, be prepared and learn from them. The flower peddler who made Dave's skin crawl sure was a salesman, as his behind-the-table persona was 180° from the pig that set up earlier that morning. The family, on the other hand, created real bonds with their clientele and other vendors, and they worked like dogs. They were awesome folks. Once, while looking over their raspberries, Dave mentioned that our canes were picked clean by the kids, but we would love to make jelly. With a smile, the owner said, "Stop by our farm on your way home." When Dave arrived, she had 20 "ugly" quarts ready for him to take home, free of charge. All he had to do was pick through them to weed out the ugliest, but most of them were ok and made great jelly. We repaid her favor with some prime mushrooms for a family gathering at a very reduced rate. These are the bonds you want. You might even find yourselves cross-promoting one another and working together to create and distribute recipes featuring both of your products. Raspberry shiitake pork glaze, anyone?

Another thing we've learned is that your tent is important. First, we bought a cheap tent (less than $100) because we didn't have the budget for a better one. One market we were looking at was in a church parking lot on top of a hill. It was along one of the main arteries for commuters, so we thought it would be a good location. It wasn't far from home and was an afternoon market, so no 4 AM wake up. We did fairly well, had a nice clientele—a number of nurses and professionals would stop on the way home. Meat, veggies, fruits, and of course our mushrooms, made for a nice layout. But, if it's too good to be true…as the old adage goes. It turned out that the location was a veritable wind tunnel. One day the sky darkened west of us, a maelstrom came along all of a sudden and shredded our tent. This happened a few more times, costing hundreds of dollars, before we had to abandon the market. Our advice, invest in sturdy tent and protect it. You want to make a profit, not be the sole sustainer of your local tent salesman.

We learned from first-hand experience (and from hearing Joel Salatin speak at a Mother Earth News Fair) that farmers markets are social gatherings. There are no grocery carts at the entrance; it's about more than getting groceries. People are going to buy what they can carry, and most people can't carry much. In short, don't inoculate 200 shiitake logs and take them to a farmers market, thinking you're going to sell all of them at $25 apiece and walk home with $5,000. You can only sell that many logs at really big events with close ties to sustainable agriculture, and even then, 200 logs is a trailer load! One or two per market is not a bad idea, to let people know you have them. They are good sellers, and if you have one "popping," you can explain how you grow them. At first, we saw farmers markets as an income producer and they were, to a certain degree. But, markets alone won't sustain you. What you really want to do by vending at farmers markets is advertise your services and products, eventually creating a depth of local consumers who return to you at the market and in other ways, such as by taking one of your classes.

Do your research and figure out at which markets you can do well. A giant farmers market may not be the best option, as you might get lost in the crowd. Most weeks, we sold just as many mushrooms at our 10-vendor Saturday market as our 100-vendor Saturday market (we split up and attended both). But, be alert. A farmers market that once was a success might not continue to work. As a prime example, we had both worked in a downtown area, right across the street from the farmers market. We would always run over and grab a few items. But, that was ten years before we were vendors there. At that time, the downtown was vibrant, with business people flurrying about the streets. When we set up our booth a decade later, a number of businesses had vacated the downtown. After only a month, we stopped attending, as there were no longer lots of professionals on lunch break. This farmers market reflected the hard times that had hit the downtown. It really saddened us to see this and to have wasted time, money, and energy trying to bring a unique product to a downtown that was no longer vibrant.

If you can, attend the farmers market before you sign up to be a vendor. If you like the market as an attendee, you will probably like it as a vendor. If you find it cluttered or unwelcoming, then you will hate vending there and will lose money. See how traffic can get in and out. Is it a pleasant experience?

Other questions you may want to answer include:
- Are the other vendors friendly?
- Is anyone else selling products similar to yours?
- What are the rules?
- Can you sell consigned items?
- Can you cook samples?
- Do you have to set up a tent?
- How far is the drive?
- How is parking?
- How is access to customers?
- What type of advertising are they doing?
- What are the costs?
- What happens to attendees if it rains? Can they hang out somewhere dry and then continue to shop or will everyone just pack up and go home?
- Does the market have volunteers to help you with setup and tear down?

The feeling you get and the answers to these questions will help you choose the right markets. Everything should be written and spelled out, but don't be surprised when you have to set up in a sub-par spot for a month or two because the market manager's nephew is now in your spot selling pottery—leaving you next to the Port-a-Potties on the hottest August days. These are things you'll want to research and consider upfront because if you leave a market mid-season you'll have probably lost the opportunity to vend at a better venue for that season. Hopefully, the lessons we have learned will help you avoid some pitfalls. If you do your research, you will be happy when 4:30 AM rolls around the morning of your market, even if there's a hanging drizzle in the air. You won't mind attending the good farmers markets, but the bad ones will have you asking, "why me?"

Joking aside, your farmers market is your chance to shine weekly, face-to-face, with local customers. Come to market on time, in presentable fashion, and make your booth attractive to mushroom lovers. For example, you could have a large, waterproof sign picturing a handsome mushroom, stating "Fresh Mushrooms." If selling grow kits, a fruiting one will attract customers. Consider featuring a weekly recipe that includes ingredients from multiple vendors at the market. This takes a little coordination and communication among vendors, but once you have some recipes for various times throughout the growing season, subsequent years should be easier. Provide samples if the market allows on-site cooking or will cook the mushrooms for you. There's no better way to market your mushrooms than by taste.

Here's a sample farmers market checklist, which can also be adapted for other types of shows:

- Mushrooms, obviously
- Fruiting mushroom grow kit or block, if you have one
- A sturdy tent
- Rope with weights to anchor your tent (on paved surfaces) or just rope and stakes for grassy areas. Hint: It's always good to ask whether you'll be setting up on grass or pavement when you attend a market or outdoor show so that you're prepared.
- Tables
- Chairs
- Cooler
- Your other products
- Recipe cards
- Cash box with plenty of change (depending on the volume sold at a market, have between $100 and $300 in change), chalk, pens, receipts, a small notebook to write down items to follow up on with customers or things to change, extras like tags, whatever other small items you need
- Smart phone or tablet with credit card swiper
- Blank receipts
- Product sales record sheet (see example in the Appendix) if not taking pre-market and post-market inventories (we found the sheet to be a lot easier, as we did not have to inventory twice for each market, and tick marks were easy to handle during the

FIGURE 9.10. One of our door prizes.

FIGURE 9.11. Table presentation with shiitake block.

market; plus you can write your starting balance, which is the amount of change you brought, and later record your profit.)

- Marketing materials, such as business cards, e-newsletter signups, on-farm workshop announcements, brochures. Hint: Don't forget racks or weights to hold these down when it gets windy. If you have on-farm or other events, it's good to have an upcoming events flyer. See example in the Appendix.
- Display boards with pictures of your products, a chalkboard listing the day's products and price
- A farm banner to hang on your tent
- A tapestry or tablecloths to dress your tables
- One helper at busier markets; a few at large shows
- Supplies to cook and provide samples, if allowed at the market
- A door prize, in exchange for exposure, if attending a large show

Special Events and Shows

Whether named "Farm-to-Table," "Farm-to-Fork," "Farm Fest," "Plow to Platter," or something else, local events that connect local farms to local consumers are on the rise. If you direct market your products, don't miss this type of event near you. Some examples of local events to find and consider attending (or creating) include:

- National Mushroom Month events (September)
- Mother Earth News Fair
- Green festivals
- County fairs
- Local harvest festivals
- Home and garden fairs
- Sustainable and country living fairs
- On-farm, multiple farm vendor events

FIGURE 9.12. Pink oyster grow kit demonstration.

FIGURE 9.13 Berglorbeer Farma display board at farmers market.

FIGURE 9.14. Aedan and Cassidy Sewak helping at our first market.

FIGURE 9.15. Cooking shiitake samples at a market.

- An event on your own farm, such as a harvest festival, customer appreciation day, classes, or school field trips
- Farm-to-table events
- Sustainable agriculture conferences
- Heritage, arts, film, or other types of community festivals
- Fundraising galas or receptions

Let's take a look at the school field trip example. Kids tell their parents how things are, so consider hosting annual field trips for local elementary schools. If you create a great experience for the students attending, their parents will hear about it. Here's an activity you could provide: have the kids set up a small-scale, eventually edible, mushroom growing experiment to take home, with notes for the teacher on how the kids could report the results to the class and to you. An online reporting system could be set up to track the kids' results. A recipe could accompany the take-home packet. The activity could be called "A Tasty Experiment."

With regard large shows, we found that any organic, farm-to-table, green festival, sustainable living/farming type worked great for us. The one massive garden show we attended was a complete flop for us, so we lost money. We had vended at an organic farm show, where we were approached by a very nice lady who ran a big garden show with huge numbers in a city two hours from our place. She had no mushrooms at the show, and she thought we would make a killing. We looked at the website, all the materials, and it really seemed like a good fit. It was expensive, so the organizer worked with us, giving us a corner booth at a reduced rate. After crunching the numbers for hotel, meals, and fuel, we decided if we sold slightly over half of our product we would break even. We took shiitake logs, oyster grow kits, dehydrated mushrooms, cooking rubs, a couple of varieties of native plants and thought we were set. We got to the facility, and after moving many wheelbarrows of shiitake logs and 100s of grow kits, we were ready to roll. The other vendors seemed very kind. There was a lot of nursery stock, garden benches, raised beds, and building supplies. It seemed to be a good enough fit.

After set up, we walked around. Beyond our neighboring two or three aisles, which had products like handmade soaps, beeswax soaps, and other natural products, we found the meat of the show: big landscaping companies, fancy outdoor living rooms, waterfalls, and $100,000 all-inclusive landscaping deals. We realized this was more of a spray-my-grass-green crowd. After three arduous days, we hadn't even sold half of our products. We would not be coming back. After talking with a number of the other "natural products" folks, it was apparent that we had been lured in to help grow and bolster the show. But the reality was that this long-running show appealed to a crowd who wasn't conscious of their landscape, they just wanted it to look good. It was the "damn those bugs, spray it greener!" crowd. Mushrooms aren't that popular in this American monoculture, "perfect" lawn society; in fact, they are viewed with as much disdain as the dandelion. But, it wasn't a total loss. We signed up a few people to attend our Shroom Classroom, folks we still talk to today, so even the worst show has that silver lining. Still, if you can scout a show before vending there, do it!

Losing money is never any fun, but we did get one of the best stories *ever* from that show. We sold a guy an oyster grow kit with instructions. We had a nice discussion. He had been a "perfectly green lawn" guy,

FIGURE 9.16. Dave at a garden show—maybe it was the look on his face that scared them away.

FIGURE 9.17. Now, that's better!

but, due to health issues he was trying to break his addiction to chemical reliance for green perfection. Dave explained to him, through photos and the instruction sheet, how oysters would pop out of the bag, how he should mist them with clean water, not city water because treated water will sometimes kill primordia. On the Monday after the show, Dave got an irate phone call from him. He was ranting about how his bag was doing nothing, that there was just white threads everywhere, and how he warned his wife to step back so she didn't get her nose broken.

FIGURE 9.18. Kristin at a native plant sale, selling mushrooms too.

When Dave asked why his wife would get her nose broken, he told him she was watching it so closely he was afraid that when the mushrooms "popped out" they might break her nose. And they say chemicals on your lawn are safe!

So let us set sail to the right shows: organic, self-reliant, farm-to-table, alternative living, back-to-the-basics shows. We found these to be the best fit, after quite a bit of trial and error. Farm-to-table shows are fun, but make sure to talk to the organizers and ask questions about how to get the best results. For one show, because we discussed things ahead of time with organizers, we knew to provide fresh mushrooms to the show, which had a culinary school on hand to cook up samples for visitors from the vendors' wares. This helped to drive sales tremendously. Anytime you can have people taste your variety of mushrooms, do it. They will sell themselves once someone tastes a fresh mushroom plate.

The pessimist complains about the wind; the optimist expects it to change; the realist adjusts the sails.—WILLIAM ARTHUR WARD

Large shows and on-farm classes have been our bulk money makers, by far. Shows that cater to a simpler life help to provide larger sums of income than farmers markets alone and had the potential to allow us to thrive, provided that we could find more such shows to attend than the few we initially did. They are hard work and are always so busy that we hardly ever get to listen to talks or mingle much with other vendors. You will be busy at such a show, and that is a good thing. Plan out who will work the booth and when breaks will be. If you can, enlist friends to help at the booth. That way, you'll be able to wander and have time to make connections and learn about complementary activities or business ideas. Just as mushrooms can become part of your life, so can beekeeping, soap making, and a whole host of other activities you may want to add to your increasingly sustainable lifestyle. Plan ahead, so you are not just a slave to the mushroom stand!

The right shows are a great way to make money, meet new people, trade ideas and goods, and learn. You have to be prepared, though, and therein lies the harsher reality and true mayhem. You will want to make sure you have enough product. Given the nature of mushrooms, this requires planning—lots and lots of planning. We usually sell all of our products, but our best sellers at big shows have been oyster grow kits and shiitake logs. Since we always had logs on the property, it was a matter of figuring out how many we could/should take, how many we would need to inoculate before/after a show—and when. Reishi logs in 3- and 5-gallon buckets were the same. The true test of your stamina and ability to be a mushroomer is to create 200+ oyster grow kits in a very short period of time. You will need to have your spawn ready to rip 5 to 6 weeks out. Optimal bags will be created 3 to 4 weeks before the show, depending on your strain, so that when customers receive their kits, they won't have to wait a month for fruiting. This is where keeping notes on your strains and their run times is so important. You don't want all your grow kits popping at the show. If you can get a couple of bags fruiting for demonstration, ideally one for each strain/species, you should sell out (same for shiitake logs and reishi logs). Large show preparation will require long hours every day for everyone involved—14 to 16 hour days aren't uncommon for our show prep. While Dave and the interns work on mushroom kits, Kristin prepares marketing materials and takes care of logistics. You must be prepared to stand out, and you want to maintain an organized booth. Remember to consider logistics for your customers. They may want to buy a couple of shiitake logs, but not carry heavy logs around for the rest of the event. Sell them the logs and put them aside. Have your customer pick them up when they are heading home. This requires planning and organization. Such is the life of a fun guy crafting fungi.

Roadside Stands

Especially because of the short shelf life of most mushrooms, the direct marketer should consider establishing a roadside stand location. This way, when a large flush of mushrooms is ready, you have the option of selling at least some of them at their optimum peak rather than having to wait until Saturday's market and thus risk not being able to sell them at all.

You-Pick Opportunities

This is where real-time social media can be super powerful. If you are willing, customers could come to your farm and pick their own mushrooms, as they are available. You could post such opportunities on your social media sites. You can control the hours you are available to host customers on-site. But, you must be willing to host customers and be prepared with a nice farm appearance, plenty of change on hand, and a friendly greeting.

Farm Store

For more sophisticated, larger farm operations, a farm store may be an option. An outbuilding can be converted into an on-farm stand with weekly, advertised open hours. However, you need the manpower, sufficient marketing and interest, and enough product to establish such a regular store operation, even a simple one. Farm stores can also be developed cooperatively, such as through the local food network.

Developing Your Unique Marketing Strategy

Dave's sister Kathy and her husband Jody of Tsoma Farms have a successful mushroom business almost 1,000 miles from ours. They are consistently able to market their mushrooms almost solely by attending farmers markets and by word of mouth. We had to get creative. Our customers were primarily found at larger events outside of our small town, and our most successful farmers market was an hour away. Almost all of our class attendees were from larger metropolitan areas nearby. We had to get out of town to be successful, because our local market was not that strong. We started to ship DIY mushroom kits to people beyond our local market.

No two communities are the same. What may work in one place could be disastrous in another. Though so many marketing possibilities exist, even beyond those included here, you cannot do it all. So, be smart in your approach. Pay attention to your unique market.

Here are some basic tenets you will want to follow:

- Create a basic marketing plan as part of your overall business strategy (see the list below).
- Get to know your unique market.
- Start small and build.
- Keep monetary marketing outlays to a minimum at first.
- Budget marketing time.
- Consider deploying marketing interns, minimizing your time and maximizing your efforts. Challenge each incoming intern to enhance one aspect of your marketing efforts and leave you and your next intern with the tools to keep it up.
- Be creative!
- Evaluate and reevaluate.

Small Farm Marketing Strategy

A Marketing Strategy

Answer the following questions to develop a simple, yet effective marketing strategy. Keep in mind that your goals and activities need to be SMART (Specific, Measurable, Attainable, Relevant, and Time Bound) and that your goals could largely be included in your overall business plan, rather than in your marketing strategy, as marketing activities will help you achieve your overall business goals. If the results are not detailed enough, you can always go back later and fill in the blanks. However, farmers cannot afford to spend tons of time planning. Therefore, this worksheet is designed to minimize the time it takes to develop a simple, but solid marketing strategy.

1 Locally, who are my potential customers?
2 Where and how can I find these local customers?

3 In a larger area (such as for the sale of prepared, non-perishable DIY mushroom growing kits), who are my potential customers?

4 Where and how can I find these customers?

5 What is my monthly marketing budget? For time? Money?

6 Based on the answers above, what are my weekly marketing tasks?

7 What additional monthly marketing tasks can I commit to accomplishing?

8 What additional annual marketing tasks can I commit to accomplishing?

9 Based on my target customer bases, what are the top three marketing strategies I should get in place and deploy upon launch of my business?

10 Based on my target customer bases, what are the top three marketing strategies I should deploy within years 1–3?

11 How am I going to connect with and continually engage my customers, transforming them to friends and ambassadors?

12 How am I going to evaluate my online, on-site, and more traditional marketing efforts (such as paid advertising)?

13 Are my marketing activities helping me to achieve my overall business growth goals? How? Which ones are working best? Which are not?

Evaluating Your Efforts

Always ask your customers where they heard about you. Ask this on your order forms, website contact forms, face-to-face, and any other avenues you can think of. This way, you will know which of your marketing strategies is working best and which you might want to focus less time on. Every local market is different, so what might work very well in one geographic area could be a waste of time in another. Here are a few basic question you should ask your customers regularly, depending on the situation:

1 How did you hear about us? (new customers)

2 What compels you to continue to do business with us? (returning customers)

3 What are we doing well?

4 What could we improve?

5 Would you recommend us to your friends?

6 If you answered "yes" above, could you provide a short (1–2 sentences) testimonial for us to use to promote our business?

7 How did you prepare and enjoy our mushrooms?

 The only constant is change.—HERACLITUS OF EPHESUS

Looking Forward and Anticipating Change

Great leaders, even of family businesses, are able to effect change *and* anticipate the emerging future, which is simply what is over the horizon. Preferences change. Markets change. Even massive changes can occur rapidly. We have seen swift and significant changes transform the businesses of two friends of ours, both in the fly fishing industry. One friend started a national fly fishing trade show around the time that the movie *A River Runs Through It* was released in theatres. His business, and fly fishing businesses everywhere, took off due to Americans' new-found love affair with fly fishing. Another friend, also in the fly fishing business, benefitted greatly when young girls started wearing dyed chicken feathers in their hair, as he already produced chicken hackle for the fly fishing business. Though these fellows found success partly by being in the right place at the right time, their stories illustrate how massively and quickly things can change. Wouldn't it be great if we could anticipate such changes that could similarly benefit our businesses? But, you ask, how? Well, though there's no magic formula, you *can* increase your chances of success by following these tenets:

- The only constant is change. Change is inevitable. It will happen, there's no doubt. So, doing your best to anticipate future change may pay off big time.
- Just as natural ecosystems are most resilient when in a state of *dynamic* (changing, pulsing) equilibrium, economies and business are too. According to the Second Law of Thermodynamics,

equilibrium (without change or evolution) is the end state, the death of a system, such as when a machine breaks down. The lesson: Do not favor balance over change, even if balance is more comfortable and you fear change. Be willing to embrace and take advantage of changes, even instability, by brainstorming ways to benefit from coming changes.

- To attempt anticipation of changes, look at current trends, such as the increasing mushroom consumption by US citizens, evolving technologies, generational communication transformations, health and nutrition research developments, unmet needs, and problems to be solved. For example, could your mushroom business evolve to supply mushrooms that would clean up hydraulic fracking wastewater, a harmful byproduct of deep shale gas extraction?
- Get creative. Is there something that people need but don't yet have? Important: If you develop a completely new product, check into whether you need to protect yourself with a patent.
- You can also effect change with intention, effort, and collaboration. For example, you can strengthen your local market by working with other producers to reach and educate consumers, such as through a local food network.

Regardless of how small or big, simple or sophisticated your mushroom business is, you will want to adapt and take advantage of future changes. You can do this by putting some strategy and creativity to work, researching trends, and by being in the right place at the right time.

SPREADING THE SPORES

Conclusion:
Spread Your Spores

It's not all about the mushrooms. But it should be *more* about the mushrooms. Why not? Society has a lot to gain by embracing mushrooms, as they possess the power to heal and balance bodies and ecosystems. They possess solutions to many of society's ailments, from cancers in our bodies to cancers on our landscapes. Many mushroom solutions are known today, yet we still have much to learn about their potential. Could mushrooms fill the niche of water and landscape restoration? Could they serve as part of a multi-pronged, healing alternative to chemotherapy? What other niches could they fill? How do they fit into our interconnected, interdependent systems? How do mushrooms enhance our quality of life, only through their umami or in a myriad of ways? Can your engagement in mushroom cultivation add yet another way that mushrooms benefit us?

Mushrooms give us :

- Umami in our meals
- Health and beauty
- Pride and joy
- Open minds
- A clean environment
- A diversity of plants
- Quality soil
- Eco-friendly packaging material

- CO_2 for the greenhouse
- A hobby
- Difficulties and challenges
- Connections with others
- Income opportunities
- Wisdom
- Inspiration
- Mayhem!

What else do they give?

In Dave's journey, which has been very different from Kristin's, mushrooms have been "just part of his fabric." His many travels harkens that Grateful Dead lyric, "What a long, strange trip it's been." Dave has hunted wild mushrooms from the Pacific to the Atlantic; they are everywhere. When living in Virginia, he could find morels fairly consistently. He recalls fishing in the George Washington National Forest and having fresh brook trout stuffed with morels, onions, and peppers over a fire. As he lay there by the fire with a full belly, he thought "I wonder what the poor folks are eating tonight," even though he was close to penniless at the time.

Out of clutter, find simplicity. From discord, find harmony. In the middle of difficulty lies opportunity.—ALBERT EINSTEIN

We have now sold Berglorbeer and moved over 2,000 miles west to Montana for yet another adventure. It was the hardest thing to do. All of our garden boxes, mushroom beds, and madman experiments are in someone else's care. The buyers were interested in the gardens, but not the mushroom projects, which is a shame and really drives home our need to always remember the kingdoms of life, and how fungi is one of them. Time and again, we talk with folks who love organic gardening and living, but do not include the fungi into their schema. We live in Montana now. Family, friends, our paradise on the mountaintop, are all in the rearview mirror. Rocky Mountain high prairie interplay is such a different realm. But, at our new home, we have water, hardwoods, and a twinkle in our eyes. We have already found boletes and oysters, and we know the morels will do their damnedest to hide from us, both the blondes and the black ones. But, they will be found, and mushrooms will be grown and integrated into our garden. This time, we'll start with wine cap stropharia and set up the outdoor grow areas, with a long view. We'll learn all the new trees and mushrooms, the rhythms of the West, and bring mushrooms to our new home. It is mycelial mayhem, a way of life.

Whether your story is more like Dave's, one of a lifetime of mushrooms, or more like Kristin's, with introduction to them in adulthood, join us in further developing mushrooms as part of your life and in sharing our collective knowledge and passion with others, so that they, too, can benefit. Whether you're growing mushrooms for your family and friends or setting out on a mushroom business mayhem (adventure), disperse your spores by sharing your mushrooms, your knowledge, and your passion. Like mycelium, spread out and connect with others and strengthen the world of mushrooms and their benefits. Become strong advocates and partners for healthy people and planet. Collaboration is the way of the future. No one can solve our problems alone. So, let's get together for these causes, forming a community that creates change. Only when this network—the mycelium—is strong enough, will change happen and we'll see the fruits of our labors. Spread the mycelial mayhem!

APPENDIX

Marketing and Communications Materials

Workshops
Shroom Classroom Sample Letter to Participants

Dear _____ ,
Thank you for signing up for Shroom Classroom, our mushroom cultivation workshop! We will provide you with a one of a kind experience that will give you the tools you need to enter into mushroom cultivation with confidence and incorporate fungi into your diet and landscape.

If you have any specific dietary needs, please let us know ahead of time so that we can prepare for that, as we will be providing breakfast and lunch to you at the workshop. We want you to be comfortable, fed, and happy! Speaking of comfort, we will be outside during a portion of the day, so you should be prepared and dress appropriately. We will have tents set up in the event of rain. If you log on or watch WJAC TV weather, www.wjactv.com/weather, prior to the workshop, please note that we are usually somewhere between 5 and 10 degrees cooler than Johnstown's temperatures.

We look forward to meeting you and showing you the wonderful world of fungi! You are welcome to take pictures during the workshop, and we will be doing the same. On the day of the workshop, we'll ask that you fill out a one page standard participation agreement.

Attachments include the workshop agenda and directions. If you have any questions, please feel free to call or email me at _____

One final item of business: Your balance, due the day of the workshop via cash, personal check, or credit card, is: $_____

Thank you for your participation!

David Sewak
Owner, Mushrooms Manager
Berlgorbeer Farma

Shroom Classroom Participant Information
Form and Workshop Agreement
The upper half is the form participants would fill out when they sign up for the workshop. The lower half attendees can fill this out ahead of time and bring it with them or fill it out on the day of the workshop.

🍄 Berglorbeer Farma 🌾

WORKSHOP PARTICIPATION AGREEMENT

Workshop: <u>Shroom Classroom</u>   **Date:** _____

Participant Information

Name of Participant_____

Address _____

Primary Phone _____   Other Phone_____

Email _____   ◯ Sign Me Up for Cap Chronicles E-News

Emergency Contact & Phone_____

Please list medical or health issues, such as allergies, dietary needs, risks, or limitations:

_____

Terms of Participation

Participant
I, _____ (participant) agree to follow all of the health, safety, and ethical rules set forth by Berglorbeer Farma. I understand that I will likely be required to participate outside in various types of weather. I also agree to hold harmless Berglorbeer Farma from any claims and/or litigation arising out of this workshop arrangement. I agree to pay, in full, any balance listed below while attending the workshop today.

_____   _____
Participant Signature   Date

Provider
We, David and Kristin Sewak, Proprietors at Berglorbeer Farma (workshop provider), agree to lead all health, safety, and ethical procedures of Berglorbeer Farma and provide participants with the best possible working conditions, guidance, and atmosphere.

_____   _____
Provider Signature   Date

WORKSHOP FEE BALANCE: $_____
Due during the workshop via cash, personal check, or major credit card.

Shroom Classroom Agenda

Sending this out to participants in their confirmation emails and then handing it to them at the beginning of the day lets them know what will be happening—which heads off lots of questions!

Berglorbeer
& Farma

Shroom Classroom

A Mushroom Cultivation Workshop

October 12, 2013

Berglorbeer Farma

Agenda

8:30-9:30 a.m.	Registration and light breakfast
9:30-10:15 a.m.	PowerPoint Presentation...Mushrooms, an Introduction: What are they, how can we cultivate them, and understand this kingdom better?
10:15-10:30 a.m.	Indoor Cultivation: A basic approach and look at the Shroom Room at Berglorbeer Farma
10:30-10:45 a.m.	Break, Q&A
10:45-Noon	The "Outdoor Mushroom Yard(s)" at Berglorbeer Farma: Looking at Wine Cap Stropharia, Elm Oyster, Piopinno, Miatake and others: How to grow them and techniques being tried at Berglorbeer Farma
Noon-1:15 p.m.	Lunch
1:15-2:45 p.m.	Oyster propagation: The basics of growing oysters—Hands on with straw bags
2:45-3:00 p.m.	Break
3:00-4:00 p.m.	Shiitake logs—Hands on "drill and fill" shiitake log preparation
4:00-5:00 p.m.	Nature walk, *optional/weather permitting*; Otherwise, a discussion of "Wild Gathering" with identification
5:00 p.m.	Wrap up, Program evaluation, Q&A, and gathering oyster bags and shiitake logs

Shroom Classroom Handout

This is information from a handout we gave to participants at our May 18, 2013 workshop. It is intended only as an example.

Mushroom Classroom

Mushroom Identification Handout

This material is intended to help you become familiar with selected mushrooms, but is **NOT** an official identification guide.

Berglorbeer Farma Cheat sheet!

Wild and Cultivated Mushrooms for Pennsylvania

Grifola frondosa (Hen of the woods, sheep head, maitake)

Wild-This excellent mushroom is found in the fall, look around oak stumps, "Hens" are very hard to see initially, but once found will usually produce every year at the same stump! Cool nights and autumn rains bring them out.

Cultivated - To date I have not been able to produce a "Hen" to maturity indoors, these mushrooms will "club" but will not frond out. Inoculating a cut oak stump with dowel, or sawdust spawn is a better method, though you will have to wait.

Hericium erinaceus (Lion's mane, pom pom, satyr's beard, bear's head, monkey face)

Wild- This delectable mushroom does grow wild in Pa. on dead or dying oak, walnut, beech, maple, and sycamore. This is a late fall fruiting mushroom, and I have found it as late as the week before thanksgiving. When **cooked** it has a hint of crab like flavor! When you find it grab it! This delicate mushroom does not last long in the woods!

Cultivated - Fairly easy to grow indoors *If you maintain your temp and humidity!* Otherwise it will deteriorate quickly. Outdoors on logs a la shiitake but with a denser inoculation rate, and a portion, about a third of the length needs to be **underground!**

Stropharia rugoso annulata (Wine cap, king stropharia, burgundy mushroom, garden giant)

Wild - This mushroom does occur in Pa. Look for it in wood shavings, sawdust piles, and in landscaping. Books say it is a spring to fall fruiting mushroom, but I have never found it before late June. When the gills begin to color it is best to leave it alone, or gather to inoculate your landscaping. It gets a bitter taste when the spore begin to mature.

Cultivated - Fairly easy, either gather wild maturing specimens and place cap down on a pile of wood chips, or purchase spawn and inoculate yourself.

Pleurotus ostreatus (Oyster mushroom)

Wild - In Pa. on numerous hardwoods, look for oyster to pop when temps fluctuate more than 15 degrees around rains, will fruit any time, but primarily a late summer into late fall/winter. Areas where logging has occurred look where one tree that has been felled has struck another tree. Also any scarred trees from wind storms, etc. An opening and oysters can appear!

Cultivated Can be grown on just about anything (except cherry trees!). Spawn can be purchased in a myriad of forms, and **inoculation** can be straw bags, mulch, trees, **rope**, etc.

Morchella (morel) *WILD ONLY* Look for it in the spring, apple, poplar, ash, hickory and white pine trees. When you see the first dandelion of the year start looking, and keep looking, this prized mushroom and its hunters keep their secrets! Old farmsteads, spring boxes have been great areas, where you find them one year they may not be the next. They appear in grey, yellow, and dark varieties in Pa, their unique shape and hollow stems give them **away**. They are difficult to spot but once one is found *usually* the forest floor comes alive with many!

Lentinula edodes (Shiitake) *Cultivated only:*

Indoors This mushroom can be grown on sawdust blocks fairly easily, with any indoor mushroom keeping the humidity and temp. stable is a must.

Outdoors Drill and fill logs are the best method, these mushrooms prefer oaks, beech, maple and other hardwoods. Freshly cut trees must be used; dormant trees are the best — late fall through spring. But trees recently deceased can be used, if using trees from the spring through fall (when trees are active) allow the logs to sit about 2–3 weeks, **but don't let them dry out!**

Craterellus cornucopiodes (Black chanterelles, horn of plenty, black trumpets)

Wild Only- These excellent edibles are easy to find, *once you find them!* For some reason they blend so well into the forest floor but once you spot one, you see them everywhere and soon you can have pounds, hence the horn of plenty moniker. I have had great luck finding them around beech trees, ***NOT beech thickets***, but where beech trees are interspersed with other hardwoods. Along stream hillsides, nestled in the moss, where decaying logs cross a beech's sucker runner! These are easily dried and stored, eaten fresh. When found, just stop, slowly and easily lower yourself down so as not to trample any more than you have! And settle in to pick, they are small but plentiful when you find them.

← **Polyporus sulphureus** (Chicken of the woods, sulpher shelf)

*Wild-*Nothing else like it, orange on top, yellow underneath. From summer into late fall after rains, simply walk through the woods looking for orange, once you found them bring bags and lots of them, young are excellent, bigger woody ones the outer edge should be trimmed off.

Flyer for Gardening with Kids Camp

Berglorbeer Farma

Gardening with Kids Mini Camp! *Grow with Us!*

What? A half-day early summer mini-camp for parents and elementary children (ages 5-12) that teaches families how to grow their own food and grow greener together!   Subjects will include: Techniques that appeal to kids, gardening basics, soil & worms, composting, "square foot" methods, good bugs, "3 sisters," green & organic practices, and native pollinator plants.

When? Sunday, June 23rd, 2013, 1-5pm

Where? Berglorbeer Farma, [ADDRESS]

Why? Gardening reconnects people to nature and wholesome food, fosters a sense of pride, builds family bonds, increases self-sufficiency, and provides control over what goes into your family's food.

Cost: $50 for one parent and one child; $10 extra for each additional family member. Includes course, materials, a vegetable plant, and food following the workshop. Discounts for military families, PASA and Western PA Mushroom Club members.

How to Register: Visit our services webpage and provide a secure online down payment of $25 or more via our PayPal system: [URL]

Questions? Email or call David Sewak: [EMAIL] or [PHONE NUMBER]

Care Sheets

Grow Your Own Oyster Mushrooms!
Instructions for how to care for your grow kit

Remember to always pick up your kit from the bottom to avoid breaking the seal at the top of the bag

Place your grow kit (also called a tabletop garden) some place inside your home out of direct sunlight. The kit does not require complete darkness, but does not flourish under direct sun. We recommend that you place it somewhere it won't be moved or disturbed, but also somewhere where you will not forget about it. For this reason, we advise against placing it in a garage or tool shed, unless that space is used almost every day. When your oyster mushrooms begin to grow from the bag, they will only take 1–3 days to reach maturity, and will quickly become past their prime if not attended. The ideal temperature range for the room where your kit is located is between 60° F and 90° F. If temperatures fall below 55° F, the kit may cease growing and may not produce mushrooms. (This applies to pink oyster mushroom tabletop gardens, while PoHu and grey dove mushrooms thrive in 55°–85° F)

When your grow kit is very white all over, it is completely colonized with the mushroom mycelium (the root structure). It may already be this way when you purchase it, but can take up to 3 weeks to reach this state. Once the bag is fully colonized, cut a small X into the plastic on each of the four vertical sides (front, back, left, and right). Keep these cuts misted with a spray bottle. Using non-chlorinated water (distilled water is best). Mist all openings at least once per day. In hotter temperatures, it is recommended that you mist at least twice a day. Your mushrooms will grow toward the oxygen, right out of these cuts.

Mushrooms are ready to harvest and cook when the caps are between the size of a quarter and the palm of your hand. Continue to mist your mushrooms until you harvest them. For best flavor, we recommend that you harvest your mushrooms before the caps become covered in a powdery substance (the spores released by mushrooms above them.) It is also recommended that you use them before the gills become yellow or the caps begin to dry out and feel leathery. You can cut the mushrooms away from the bag with a knife or kitchen scissors, or simply pull gently on the cluster until it comes away from the bag. Remove tough stems and straw or cottonseed hull material prior to cooking.

Store mushrooms in a brown paper bag in the refrigerator for up to one week. To retain moisture and extend freshness, fold the top of the paper bag over a few times. If you cannot cook all the mushrooms your bag produces, you can dehydrate them or freeze them. In this case, you will need to rehydrate your mushrooms before cooking in boiling water, stock, or near boiling cream.

After your kit fruits (produces a flush of mushrooms) it will rest for 2-4 weeks (depending on the strain) and then produce new mushrooms. You may tape over your original cuts to retain moisture. We recommend using clear tape so that you can see if mushrooms begin to grow beneath your original cuts. It is also

Berglorbeer Farma

Berglorbeer Farma 🌸

recommended that you gently squeeze excess air out of your bag before re-sealing any cuts. You will know your kit is ready to fruit again when it begins to "pin" or form tiny branches that will become mushrooms (see image below.) They are often very small, so look all over your kit very closely once in a while after removing the first flush of mushrooms. When this happens, carefully cut a new X into the plastic directly over the pins and keep the cuts misted. This will cause the pins to grow out of the bag into full4 sized mushrooms. Continue this process until your kit ceases to produce mushrooms. When the kit does stop fruiting, feel free to compost the contents, or break it apart and use it as mulch in your garden!

Summary
1. Place bag in a stable location, away from direct sunlight, in 60-90º F temperatures.
2. Watch for full colonization. When kit is fully colonized, cut an X into each of the four sides.
3. Keep cuts misted and watch for mushrooms.
4. When mushrooms begin to grow, harvest within 2-4 days.
5. Store harvested mushrooms in a folded paper bag, in the refrigerator, for up to one week.
6. Save excess mushrooms by dehydrating or freezing and rehydrating as needed.
7. After harvesting your mushrooms, you may tape over your cuts with clear tape, squeezing out excess air from bag in the process. If you do *not* tape over these cuts, continue to mist them once a day.
8. After 3-4 weeks, your bag should fruit again. You will know because of the growth of mushroom pins.
9. Carefully cut holes above pins, and keep misted until you harvest your next batch of mushrooms.
10. Continue to care for the kit as described above until bag ceases producing all together.
11. Once bag is spent (finishes producing for good) compost or use as garden mulch, or spread in a moist, shady area (if outdoor temps are high enough) and watch for any remaining flushes. Keep moist.
12. Enjoy!

Mushroom pins forming inside the bag

Berglorbeer Farma

🍄 Berglorbeer Farma 🌿

Reishi (*Ganoderma resinaceum*)

Care

You are on your way to growing you own medicinal Reishi mushrooms! This pony log has been inoculated with Reishi spawn. Simply place in a shady, warm location where you can watch them. Keep the sand moist. Under greenhouse benches are an excellent place for them, as the heat and moisture will help the spawn run! When the reishi fruits, it will have a lacquered reddish brown appearance with a white to yellow leading edge. As long as the outer edge is white or yellow your Reishi is still growing! To harvest, simply break off at the log, simmer in water for 2 hours and sip your way to a healthy tea! One fully developed conk makes approximately one gallon of tea. You can insert your log into a stump, and possibly keep your Reishi going for some years!

Berglorbeer
Farma

Rope Spawn Instructions

    The rope you have purchased has been inoculated with *Pleurotus ostreatus* the strain of which is the Grey Dove. These delicate, delicious and lovely silver blue mushrooms will fruit from late summer until freeze up in the fall. This method is for the landowner who has to harvest trees, whether for fire wood, lightning strikes or landscaping, trees sometimes must fall, why not get something from the stump? See Table for Tree species suitable for inoculation:

Step 1:

When cutting your tree leave enough of a stump sticking up that you can work safely and you have enough stump to produce significant mushrooms. Girdle the tree with a chainsaw between 4 to 12 inches above the ground-*Try not to cut too deeply or too Shallow with the chainsaw, just deep enough that the rope doesn't Stick out or is way back in the heartwood!*

Bigtooth Aspen (Populus grandidentata)
Trembling Aspen (P. tremuloides)
Cottonwood (P. deltoids and other spp.)
Balsam Poplar (P. balsamifera)
Tulip or Yellow Poplar (Liriodendron tulipifera)
Box Elder (Acer negundo)
Elm (Ulmus spp)
Tree of Heaven (Ailanthus altissima)
Buckeye (Aesculus glabra)

Step 2:

Tap your rope into the groove you have made, use a rubber mallet or simply push in with your fingers.

Step 3:

Covering with burlap (doubled over) or waxing the rope will help retain moisture; keep mealy bugs away (waxing) bolstering the mushrooms' chances to colonize the stump

Basic notes and ideas:

Use on only freshly cut trees, not some old stump that was cut years ago-it's already colonized by some other mushrooms. The taller your stump the move substrate you have and can inoculate at a higher rate. I like to put two rope strands on the stump the first one 4" from the ground and the second 6" higher up-this speed up the spawn run on the stump. During dry periods moisten the stump, as the first 6 months are critical. After that the mushrooms and mother nature will do the rest. ***DO NOT*** use treated water, rain water from a barrel, stream, spring, pond water is fine, treated water can stunt and weaken mushrooms, and the chemicals get concentrated in the mushrooms!

Sign up for *Cap Chronicles* E-News: Like us on Facebook:

 info@berglorbeerfarma.com   www.berglorbeerfarma.com

Berglorbeer Farma

Shiitake Log Care

Congratulations! You are on your way to the exciting world of growing your own mushrooms! Few things are as gratifying as fresh shiitakes that you have grown yourself. Though this log is inoculated and should produce for you, you will have to tend to your log. If you plan to "get it and forget it" you will probably have some mushrooms produce, but with a little care, you will have a number of flushes before this log is spent, which is a better return on your investment. If you follow the steps below you should be happy with the log's production! I have done my best to ensure your log produces quality shiitakes. The rest is up to you and Mother Nature!

Step 1: When you get home find a nice shady place you can put our log (Under an evergreen such as Hemlock, Pine tree, or even a robust Rhododendron)

Step 2: Get your log off the ground! Competing fungi will attack a log on the ground and ruin your log before it's time for fruiting. Also, air circulation is critical to a good spawn run. Securely place your log on two bricks, a couple of stones or some pavers.

Step 3: Keep the log moist. Mother Nature provides enough moisture in the Northeast to fruit shiitake "naturally" in the spring and fall, but there are times when we have prolonged periods of dry weather. So during these dry spells, mist with a garden hose, and even wrap it in burlap (something that will hold some moisture but breath too). As a general rule, if it hasn't rained or had a heavy dew in a week, water the log. Cracks on the end are a sure sign of drying out!

Step 4: Learn to know what a log looks like when it is ready to fruit. See photo below:

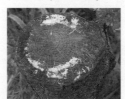

Step 5: *Fruiting the log:* Your log should fruit naturally this spring, but due to localized conditions this is not always the case. If May arrives and you don't see any pin development but notice that the spawn run from step 4 has occurred, you can force a fruiting. Simply submerse your log in cold water (A clean garbage can work-just scrub out with diluted bleach and rinse thoroughly) for 24 hours. When you remove the log, drop it on its end or wrap the end with a hammer. This "soak and strike" method will prompt a log to fruit. Stand your log upright (off the ground) leaning against a shady tree. If the log starts fruiting during a dry period, mist your log twice a day, as this will improve the mushroom quality and yield! Force-fruiting may cause your log to be spent sooner, but you will have a more consistent supply of mushrooms, and *you* control when you have the supply. Without forced-fruiting, logs will produce in the spring and fall for a couple of years, with the supply getting smaller as time goes by.

Step 6: *Resting your log:* After each fruiting, allow your log to recuperate for about 60 days. To do this, put the log back where you had it originally, but still keep it moist.

Step 7: *Enjoy!* Shiitakes are delicious and can be a very healthy part of any diet, plus it is fulfilling to watch your own log produce the best shiitakes ever!

Sign up for *Cap Chronicles* E-News:　　　　Like us on Facebook:

Berglorbeer Farma　　　　　　　　　　　　　　　　davidsewak@gmail.com

www.berglorbeerfarma.com

Price List

This was our list in September 2012. It served double duty as a record of sales.

Sample Price List & Sales Record

9.22.12

Pohu Oyster Home Kit: $20_____

Grey Dove Oyster Home Kit: $20_____

Pink Oyster Home Kit: $20_____

Shiitake Log Home Kit: $20+ _____
(price varies by size, each individually marked)

Cooking Rubs: $3 or 2 for $5_____
Deyhdrated Black Chanterelles: $3 or 2 for $5_____
(customer can mix and match rubs & chanterelles)

Photo Prints: $12 or 2 for $20_____

Native Landscape Plants in gallon pots – ON SALE: all $5.50, 2 for $9

Anise Hyssop_____

Butterfly Milkweed_____

Cardinal Flower_____

Great Blue Lobelia_____

Labrador Violet_____

Nodding Onion_____

Sensitive Fern_____

Wild Bergamot_____

Total sales = $_____

Listing for Intern

This is the job announcement and position description we used to recruit our seasonal farm interns.

Berglorbeer Farma

Sustainable Agriculture Internships 2013

Summary

Are you an agriculture, environmental studies, sustainability, or biology undergrad looking for a meaningful summer 2013 internship? Consider interning at Berglorbeer Farma! We currently have full-time and part-time positions available, which are designed to be intensely educational, hands-on, and fun.

About Berglorbeer Farma

Founded in 2012 by husband and wife team David and Kristin Sewak, Berglorbeer Farma is a family-owned, sustainable farming operation on 3 acres atop the Eastern Continental Divide near Windber, PA, intended to help people *Eat Green, Save Green, & Live Green*. We offer the highest quality, organically-grown gourmet and medicinal mushrooms, heirloom vegetables, native plants, rain barrels, Shroom Classrooms, cookbooks & other publications, and sustainable design. Our website is: _____. Our qualifications are located on the *About Us* page. We are a fully insured and licensed LLC in PA.

Intent of Positions

The intent of our Sustainable Agriculture Internships is to provide students with a wealth of experience and hands-on education regarding the business of sustainable farming, while assisting Berglorbeer Farma in growing our produce and our operation.

Learning Objectives and Duties

Interns will assist owners with many of the following duties, learning many meaningful concepts through work:
1. Gourmet and medicinal mushroom propagation, including, but not limited to:
   a. Shiitake log inoculation and care
   b. Possibly: Production of mushroom spawn using small, on-site lab
   c. Inoculation and care of a variety of oyster straw bags
   d. Inoculation of vegetable and native plants with wine cap stropharia and elm oyster, a unique companion planting designed to enrich the plant's soil and provide shade for the mushrooms
   e. Wild edible mushroom forages
2. Property design and management related to sustainable agriculture, permaculture, horticulture, ecology, forest gardens, edible landscapes, market gardens, mushroom grow spaces, native wildlife habitat, pollination, natural pest control, and beautification
   a. Assistance with finalization of site design (contribution to design and implementation through intern's creative visioning)
   b. Implementation of slated 2013 design elements, such as simple construction of additional integrated grow spaces, fencing, plantings, rain barrel and rain garden installations, and other simple landscape changes
3. Implementation of a study designed to find an effective set of organic solutions to tomato blight

𝕭erglorbeer 𝕱arma

4. Propagation, care, and management of our native plant nursery, including, but not limited to:
   a. Collection of wild native seeds, seed stratification & germination methods, and seeding flats
   b. Transplanting plugs and seedlings into larger containers for sale
5. Marketing, including, but not limited to:
   a. Research and design regarding one themed garden design pamphlet OR assistance with cookbook production
   b. New product materials: writing and design
   c. Website updates, performed via Intuit templates, *Cap Chronicles* e-newsletter, Facebook posts, and Twitter account set-up
6. Rain barrel construction
7. Chicken flock care and egg collection
8. Assistance with sale of products at farmers markets, trade shows, and other events
9. An introduction to integrative, sustainable farm management, business planning, event planning, finances, sales, social media, customer relations, and marketing
10. Other duties, as agreed upon by both parties

To be successful, an individual must be willing and able to perform each function satisfactorily and safely, with appropriate guidance and supervision from Berglorbeer Farma owners. To the best of our ability, we will work with our summer interns to tailor experiences and divide tasks based on each intern's major or areas of interest.

Period and Hours Required of Internships

Berglorbeer is flexible regarding start and end dates of each position and will do our best to accommodate an intern's spring and fall semester schedules, as well as any planned vacations. We would like each intern to start between mid-May and early June and finish during the second half of August. Total minimum hours required of part-time interns = 200, while total minimum hours required of full-time interns = 400. It is advised that selected interns attend Berglorbeer Farma's May 18th Shroom Classroom, a mushroom growing workshop for the public, as part of their orientation.

Intern Qualifications

The ideal candidate should:
- Be an agriculture, environmental studies, sustainability, biology, or other undergraduate student
- Live close enough to Berglorbeer Farma to contribute at least 18 hours per week at the farm for part-time positions or 36 hours per week for full-time positions (interns are welcomed to camp on-site on nights before work days, but must provide their own tents and bedding)
- Be willing and able to pass a criminal background check
- Be ethical, honest, trustworthy, hard-working, possess the desire to learn, and be a steward of the Earth
- Have daily access to a reliable personal vehicle that can be utilized for work, including limited travel
- Be willing to work variable, sometimes long hours, with some evening and weekend work

𝕭𝖊𝖗𝖌𝖑𝖔𝖗𝖇𝖊𝖊𝖗 𝕱𝖆𝖗𝖒𝖆

- Be able to safely lift 40 or 50 pounds, work outdoors in all weather conditions, operate simple power tools such as drills, and abide by our worker safety and food safety rules
- Be willing to commit to the position for the entire time period
- Have personal health insurance

Compensation and Credits

We have lined up a host of benefits for our interns, including:

- College internship credits at partnering collegiate institutions (credits allotted and requirements may vary; contact your advisor or Berglorbeer Farma to ensure availability of credits)
- A weekly harvest allotment and daily food tastings for each intern
- Discounts on Berglorbeer products and services for interns and their immediate families during internship duration
- End-of-year gifts of appreciation, including plants and mushroom tabletop gardens
- Lunch on all full (7+ hr.) work days
- On-site wooded or grassy lawn camping accommodations and daytime access to the appropriate facilities on-site, including restrooms, a laptop computer, a wide array of educational resources, kitchen, and outbuildings
- Mileage reimbursement for any trips taken with personal vehicles, beyond the farm
- A wealth of knowledge and experience with major aspects of a sustainable farming business
- Internship course materials to take home
- Opportunity to participate in high quality events and meet many contacts in field of sustainable farming, including a chance to return for a day or two in the fall to assist at Berglorbeer's Harvest Fest (usually early October, date TBD) or Mother Earth News Fair, occurring September 20–22, 2013, Seven Springs Mountain Resort (tickets, accommodations, food, and product will be provided as compensation)
- Provided that the intern proves to be dependable and performs well, Berglorbeer would be happy to serve as a reference for or provide letters of recommendation to the intern following the internship

How to Apply and Questions

Feel free to contact us with questions via email (below) or phone   To apply, send brief cover letter, including at least one reference with contact information and your primary area(s) of interest relevant to our internships, as well as your resume, with relevant coursework completed to-date, by April 30th to:

Kristin Sewak, Proprietor
Berglorbeer Farma

OR email to Kristin:

North American Resources by Subject

This list is not meant to be comprehensive, nor timeless. In advance, we apologize to anyone or any company that is not listed here, but probably should be. Listings are intended to provide starting points for beginning growers. As a serious mushroom hobbyist or purveyor, you will need to stay current on the world of mushrooms. Explore these and other resources for mushroom cultivation advances, up-to-date nutritional discoveries, events, and other information. To suggest additional resources for our website or future book editions, visit mycelialmayhem.com. Listing here does not constitute endorsement.

Agriculture
Agriculture and Agri-Food Canada
    AgPal: Programs and services for Canadian farm businesses, agpal.ca
Resilient Agriculture: Cultivating Food Systems for a Changing Climate
    Laura Lengnick
Secretariat of Agriculture, Livestock, Rural Development, Fisheries and
    Food (Mexico): sagarpa.gob.mx
*The Emergent Agriculture: Farming, Sustainability and the Return of the Local
    Economy*: Gary S. Kleppel
US Department of Agriculture (USDA): usda.gov

Associations
American Mushroom Institute: americanmushroom.org
Mushroom Council (USA): mushroomcouncil.org; mushroominfo.com
North American Mycological Association (NAMA): namyco.org
Organic Trade Association: ota.com
Sociedad Mexicana de Micologia: sociedadmexicanademicologia.org.mx/
Weston A. Price Foundation: westonaprice.org

Business

America's Small Business Development Center (SBDC) Network (USA): americassbdc.org

Canada Business Network: canadabusiness.ca

Canada Revenue Agency—Small Business and Self-Employed: cra-arc.gc.ca/selfemployed/

Green America: greenamerica.org

SCORE: score.org

US Small Business Administration: sba.gov

US Chamber of Commerce and Your Local Chamber: uschamber.com

Cultivation

"Financial Analysis of Shiitake Mushroom Production": M. J. Baughman

Growing Gourmet and Medicinal Mushrooms: Paul Stamets

"Growing Shiitake Mushrooms in a Continental Climate": M. E. Kozak and J. Krawczyk

Mushroom Integrated Pest Management Handbook: PA IPM Program and American Mushroom Institute

Mycelium Running: Paul Stamets

North American Mycological Association (NAMA): Mushroom Cultivation Links: namyco.org/mushroom_cultivation_resources.php

Organic Mushroom Farming and Mycoremediation: Simple to Advanced and Experimental Techniques for Indoor and Outdoor Cultivation: Tradd Cotter

Shiitake Growers Handbook: Paul Przybylowicz and John Donoghue

The Mushroom Cultivator: Paul Stamets and J. S. Chilton

Events

Green Festival: greenfestivals.org

Mother Earth News Fair: motherearthnews.com/fair

Telluride Mushroom Festival, Telluride, CO: telluridemushroomfest.org

Wild Mushroom Hunting

General

Fat of the Land: Adventures of a 21st Century Forager: Langdon Cook

"In Search of the Holey Veil" DVD: Taylor F. Lockwood

The Mushroom Hunters: On the Trail of an Underground America: Langdon Cook

Mushroom Identification Guides

100 Edible Mushrooms: Michael Kuo

All That the Rain Promises and More: A Hip Pocket Guide to Western Mushrooms: David Arora

Common Florida Mushrooms: James W. Kimbrough

Common Mushrooms of the Northwest: J. Duane Sept

The Complete Mushroom Hunter: Gary Lincoff

Edible and Medicinal Mushrooms of New England and Eastern Canada: David L. Spahr

Edible Wild Mushrooms of Illinois & Surrounding States: Joe McFarland and Gregory M. Mueller

Edible Wild Mushrooms of North America: David W. Fischer and Alan E. Bessette

Field Guide to Edible Mushrooms of the Pacific Northwest: Daniel Winkler

Field Guide to Wild Mushrooms of Pennsylvania and the Mid-Atlantic: Bill Russell

Key to Major Groups of Mushrooms: mushroomexpert.com, Michael Kuo

"Know Your Mushrooms" Video: Ron Mann

Los Hongos De El Eden (Mexico): Gastón Guzmán

Morels: Michael Kuo

Mushrooming with Confidence: Alexander Schwab

Mushrooms and Other Fungi of North America: Roger Phillips

Mushrooms Demystified: David Arora

"Mushrooms of America" DVD: Taylor F. Lockwood

Mushrooms of Cape Cod and the National Seashore: Arleen and Alan Bessette, and William J. Neill

Mushrooms of North America: Orson K. Miller, Jr.

Mushrooms of the Pacific Northwest: Steve Trudell and Joe Ammirati

Mushrooms of the Southeastern United States: Alan E. Bessette, William C. Roody, Arleen R. Bessette and Dail L. Dunaway

National Audubon Society Field Guide to Mushrooms: North America: Gary H. Lincoff

The New Savory Wild Mushroom (West Coast): Margaret McKenny and Daniel Stuntz

North American Boletes: A Color Guide to the Fleshy Pored Mushrooms: Alan E. Bessette, William C. Roody, and Arleen R. Bessette

Simon & Schuster's Guide to Mushrooms: Gary Lincoff

Wild Edible Mushrooms: Hope Miller

Wild Mushrooms of North America and Europe: Roger Phillips and Glen
Byram (iPhone and iPad app)

Trees
Fandex Family Field Guides: Trees: Steven M. L. Aronson
National Wildlife Federation's Field Guide to Trees of North America: Bruce
Kershner, Daniel Mathews, Gil Nelson, and Richard Spellenberg
*National Audubon Society's Field Guide to North American Trees: Eastern
Region*
*National Audubon Society's Field Guide to North American Trees: Western
Region*

Financing
FarmAid's Funding Opportunities: farmaid.org/site/c.qlI5IhNVJsE/b.606
0101/k.6E56/Funding_Opportunities.htm
Organic Farming Research Foundation (OFRF): ofrf.org
Sustainable Agriculture Research and Education (SARE): sare.org
USA Federal Grants Database: grants.gov
The Crowdfunding Centre: thecrowdfundingcentre.com

Local and Sustainable Food Initiatives
ATTRA Directory of Local Food Directories: attra.ncat.org/attra-pub/local
_food/search.php
Buy Fresh Buy Local/Food Routes Network: foodroutes.org
Farm Match: farmmatch.com
Farmers Web: farmersweb.com
Food Hub: food-hub.org
Local Harvest: localharvest.org
National Good Food Network: ngfn.org
Rural Bounty: ruralbounty.com

Markets and Marketing
Farm Marketing Solutions: farmmarketingsolutions.com
North American Farmers' Direct Marketing Association: farmersinspired
.com
Small Farm Central: smallfarmcentral.com
USDA Agricultural Marketing Service: ams.usda.gov
Website Builder Expert: websitebuilderexpert.com

Nutritional and Medicinal Mushrooms

Medicinal Mushrooms: A Clinical Guide: Martin Powell
Medicinal Mushrooms: An Exploration of Tradition, Healing and Culture:
　　Christopher Hobbs
Medicinal Mushrooms: The Essential Guide: Martin Powell
Mushrooms as Health Foods: Kisaku Mori
Mycelium Running: Paul Stamets
MycoMedicinals®: An Informational Treatise on Mushrooms: Paul Stamets
Shiitake: The Healing Mushroom: Kenneth Jones
Spontaneous Healing: Andrew Weil
*The Fungal Pharmacy: The Complete Guide to Medicinal Mushrooms and
　　Lichens of North America*: Robert Rogers

Publications

Fungi: fungimag.com
International Journal of Medicinal Mushrooms: begellhouse.com/journals
　　/medicinal-mushrooms.html
Mother Earth News: motherearthnews.com
Mushroom: The Journal of Wild Mushrooming: mushroomthejournal.com
The Mushroom Growers' Newsletter: mushroomcompany.com

Spawn and Other Supplies

Field & Forest Products: Peshtigo, WI, fieldforest.net
Fungi Perfecti: Olympia, WA, fungi.com
Mushroom Mountain: Greenville, SC, mushroommountain.com
Mushroompeople: Summertown, TN, mushroompeople.com
Myco Supply: Pittsburgh, PA, mycosupply.com
Penn State Mushroom Spawn Lab: State College, PA, plantpath.psu.edu
　　/facilities/mushroom/cultures-spawn
Penn State University Extension Directory of USA Commercial Spawn
　　Suppliers: State College, PA, extension.psu.edu/plants/vegetable-fruit
　　/mushrooms
Sharondale Mushroom Farm: Cismont, VA, sharondalefarm.com

Sustainability and Permaculture

ATTRA Sustainable Agriculture: attra.ncat.org
Gaia's Garden: A Guide to Home-Scale Permaculture: Toby Hemenway
Mycelium Running: Paul Stamets

Mycoremediation: Fungal Bioremediation: Harbhajan Singh

Permaculture: A Designers' Manual: Bill Mollison

Permaculture: Principles and Pathways Beyond Sustainability: David Holmgren

Sepp Holzer's Permaculture: A Practical Guide to Small-Scale, Integrative Farming and Gardening: Sepp Holzer

The Resilient Farm and Homestead: An Innovative Permaculture and Whole Systems Design Approach: Ben Falk

References

Chapter 2

Cavalier-Smith, T. (1993). Kingdom protozoa and its 18 phyla. *Microbiological Reviews* 57 (4): 953–994.

Cavalier-Smith, T. (1998). A revised six-kingdom system of life. *Biological Reviews* 73 (03): 203–66.

FC&A Medical Publishing. (2004). *Eat and Heal.* Frank W. Cawood and Associates, Inc.

Lincoff, G. (2015). *Connecting the Dots: Mushrooms and Ecosytems.* Retrieved from: garylincoff.com.

Stamets, P. and Chilton, J. S. (1983). *The Mushroom Cultivator: A Practical Guide to Growing Mushrooms at Home.* Olympia, WA: Agarikon Press.

Whittaker, R. H. (1969). New concepts of kingdoms or organisms. Evolutionary relations are better represented by new classifications than by the traditional two kingdoms. *Science* 163 (3863): 150–60.

Woese, C. R. and Fox, G. E. (1977). Phylogenetic structure of the prokaryotic domain: the primary kingdoms. *Proceedings of the National Academy of Sciences of the United States of America* 74 (11): 5088–90.

Chapter 5

Benyus, J. (1997). *Biomimicry: Innovation Inspired by Nature.* New York: HarperCollins.

Holmgren, D. (2002). *Permaculture: Principles and Pathways Beyond Sustainability* Australia: Holmgren Design Services.

Holzer, S. (2004). *Sepp Holzer's Permaculture: A Practical Guide to Small-Scale, Integrative Farming and Gardening.* White River Junction, VT: Chelsea Green Publishing.

Hutchins, G. (2013). *The Nature of Business: Redesign for Resilience.* Gabriola Island, BC, Canada: New Society Publishers.

Mollison, B. (1991). *Introduction to Permaculture*. Tasmania, Australia: Tagari.

Silverstein, S. (1964). *The Giving Tree*. New York: Harper & Row.

Stamets, P. (2000). *Growing Gourmet and Medicinal Mushrooms*. New York: Ten Speed Press.

Chapter 6

FC&A Medical Publishing. (2004). *Eat and Heal*. Frank W. Cawood and Associates, Inc.

Kuninaka A. (1960). Studies on taste of ribonucleic acid derivatives. *Journal of the Agricultural Chemical Society of Japan* 34: 487–492.

Ikeda, Kikunae. (1909). New Seasonings [Japan]. *Journal of the Chemical Society of Tokyo* 30: 820–836.

Chapter 7

FC&A Medical Publishing. (2004). *Eat and Heal*. Frank W. Cawood and Associates, Inc.

Jackson, P. (2012). The Hobbit: An Unexpected Journey [Motion picture]. Warner Brothers Pictures.

Mori, K. (1974). *Mushrooms as Health Foods*. Japan Publications: Tokyo.

Stamets, P. (2005). *Mycelium Running*. Berkeley, CA: Ten Speed Press.

Wasser, S. (2010). Medicinal Mushroom Science: History, Current Status, Future Trends, and Unsolved Problems. *International Journal of Medicinal Mushrooms* v12.i1.10.

Watson, B. (2002). *Renew Your Life: Improved Digestion and Detoxification*. Clearwater, FL: Renew Life Press.

Chapter 8

Covey, S. R., Merrill, A. R., and Merrill, R. R. (1995). *First Things First*. New York: Fireside.

Farmon.com. (2015). Quote retrieved from farmon.com.

USDA Economic Research Service (ERS) (2013). Food Availability Per Capita: Mushrooms. Retrieved from: ers.usda.gov

Venn, J. (2015). *Survey of Mushroom Growers*.

Chapter 9

National Good Food Network (2015). Field Guide to the New American Foodshed. Retrieved from: foodshedguide.org/foodshed/

USDA Agricultural Marketing Service (2012). Regional Food Hub Resource Guide. Retrieved from: http: ams.usda.gov

USDA Agricultural Marketing Service (2015). Agritourism. Retrieved from: agmrc.org

USDA Alternative Farming Systems Information Center (2014). *Community Supported Agriculture*. Retrieved from: nal.usda.gov

Index

About the Authors

DAVID SEWAK has worked in conservation and recreation for nearly two decades. First with his Papap, he has gathered wild mushrooms since he was a child and has cultivated mushrooms for over 20 years. Prior to moving from Pennsylvania to Montana, David and his wife Kristin co-owned Berglorbeer Farma, which specialized in edible mushrooms, heirloom vegetables, native landscape plants, and sustainable landscape design. David and Kristin speak on mushroom growing and other sustainable living topics at regional and national green living events. David holds a degree in History and Philosophy and is a fly fishing guide on the Missouri River in western Montana.

KRISTIN SEWAK is the founder of Natural Biodiversity, a nonprofit organization dedicated to restoring biodiversity within landscapes and involving people in environmental solutions. Prior to moving from Pennsylvania to Montana, Kristin and her husband David co-owned Berglorbeer Farma, which specialized in edible mushrooms, heirloom vegetables, native landscape plants, and sustainable landscape design. Kristin and David speak on mushroom growing and other sustainable living topics at regional and national green living events. Kristin holds a degree in Ecology, a Master of Science in Organizational Leadership, and has developed a nature-inspired leadership model.

If you have enjoyed *Mycelial Mayhem* you might also enjoy other

Books to Build a New Society

Our books provide positive solutions for people who
want to make a difference. We specialize in:

Climate Change ◆ Conscious Community

Conservation & Ecology ◆ Cultural Critique

Education & Parenting ◆ Energy ◆ Food & Gardening

Health & Wellness ◆ Modern Homesteading & Farming

New Economies ◆ Progressive Leadership ◆ Resilience

Social Responsibility ◆ Sustainable Building & Design

New Society Publishers
ENVIRONMENTAL BENEFITS STATEMENT

New Society Publishers has chosen to produce this book on recycled paper made
with 100% post consumer waste, processed chlorine free, and old growth free.

For every 5,000 books printed, New Society saves the following resources:[1]

37	Trees
3,363	Pounds of Solid Waste
3,700	Gallons of Water
4,826	Kilowatt Hours of Electricity
6,113	Pounds of Greenhouse Gases
26	Pounds of HAPs, VOCs, and AOX Combined
9	Cubic Yards of Landfill Space

[1]Environmental benefits are calculated based on research done by the Environmental Defense Fund and
other members of the Paper Task Force who study the environmental impacts of the paper industry.

For a full list of NSP's titles, please call 1-800-567-6772 or check out our web site at:

www.newsociety.com

new society
PUBLISHERS